Physics, Fun, and Beyond

Electrifying Projects
and Inventions
from Recycled
and Low-Cost Materials

Eduardo de Campos Valadares

Translated by Michael Hugh Knowles,
Heather Jean Blakemore,
and Eduardo de Campos Valadares

Upper Saddle River, NJ • Boston • Indianapolis • San Francisco
New York • Toronto • Montreal • London • Munich • Paris •
Madrid • Capetown • Sydney • Tokyo • Singapore • Mexico City

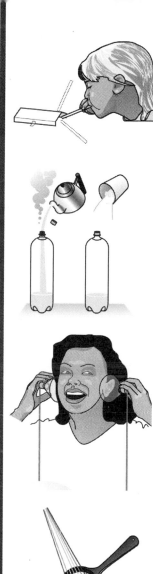

The publisher offers excellent discounts on this book when ordered in quantity for bulk purchases or special sales, which may include electronic versions and/or custom covers and content particular to your business, training goals, marketing focus, and branding interests. For more information, please contact:

  U. S. Corporate and Government Sales
  (800) 382-3419
  corpsales@pearsontechgroup.com

For sales outside the U. S., please contact:

  International Sales
  international@pearsoned.com

Visit us on the Web: www.prenhallprofessional.com

Library of Congress Catalog Number: 2005925718

ISBN 0-13-185673-1
Text printed in the United States at C.J. Krehbiel in Cincinnati, Ohio.
Second printing December 2006.

To my mother Alzira and to my father Sindulfo
To Sofia Carolina, Elisa, Felix, and Friederike

Contents

Fun with Mechanics

Playing With Light: Optics

The World of Atoms and Our World: Cold, Heat, and Giant Bubbles

Playing With Sounds: Acoustics

Electrifying Experiments: Electricity and Magnetism

Preface

WHAT THIS BOOK IS ABOUT

Physics, Fun, and Beyond is an attempt to link science education with discovery and innovation. Its basic idea is to provide a framework for further development, starting with mostly simple and inexpensive do-it-yourself projects. The suggested experiments and prototypes are a guide into "worlds within worlds," using recycled and low-cost materials, which you can find in drugstores, hardware stores, and other familiar locations. By becoming familiar with household tools—and taking appropriate safety measures—you can perform wonders. You can greatly improve your perception of the world all around you, discover firsthand that many of Nature's laws are just around the corner, and explore how they are interrelated. You will be challenged to think about all the possible applications of what you learn and to build more sophisticated prototypes on your own. Yes, you should explore all the possibilities within your reach. The basic requirements are the desire to enjoy yourself and a good dose of creativity—something everyone has plenty of. There are experiments for all tastes and ages. The proposed projects are only a few of the possible ways to discover, explore, and demonstrate how science and Nature work together. As you build on the concepts presented in this book, you will soon realize that you are surrounded by opportunities for discovery and innovation.

Another major goal of this book is to foster teamwork. Feel free to invite your classmates and friends to take part in the adventure of discovering Nature. Experience the fun of sharing your discoveries. You will learn that more ideas flow and results are attained much faster when you work as a team. The point is to find simple and inexpensive solutions. The world is looking for new discoveries. Why not display your experiments in parks, shopping malls, youth centers, children's parties, or at schools? That is the big challenge. Science is contagious. Just give it a try!

HOW THIS BOOK IS ORGANIZED

Physics, Fun, and Beyond is divided into five major blocks of experiments under the headings *Fun With Mechanics*; *Playing With Light: Optics*; *The World of Atoms and Our World: Cold, Heat, and Giant Bubbles*; *Playing With Sounds: Acoustics*; and *Electrifying Experiments: Electricity and Magnetism*. This does not mean that the experiments are isolated from each other. Quite the contrary—there are multiple crosslinks connecting the various experiments and prototypes.

The traditional divisions of physics into separate areas is totally artificial. More and more, the boundaries within physics and within science in general are being pushed further or are simply vanishing. In the twenty-first century, perhaps we will learn to see science without boundaries. Hopefully, we will learn to see the world in the same way—without boundaries—so that all of humankind can enjoy living in it and appreciate its beauty.

WHO THIS BOOK IS FOR

This book is addressed to a wide readership. Teachers and students are not the only people who get interested in and involved with science and technology. My experience has shown that curious people of all ages become thrilled with the unique way of approaching the world that science enables. This book uses an accessible language and means to describe Nature, in "as simple a way as possible, though not simpler," as once stated by Albert Einstein.

Parents with much younger children who are interested in science—or even just in "tricks"—will find here many opportunities to enjoy themselves, with child supervision and assistance, of course! There is magic in science that we can all enjoy—especially when we share it!

Most of the projects suggested are simple enough for middle school children (ages 10 to 14) or home schooling at the same level or even younger children. My experience has shown that an instructor (teacher or parent) is necessary in these cases to supervise the kids and provide assistance whenever necessary. The handling of tools, like an electric drill or a jigsaw, should in most cases be done by an adult or with teacher supervision.

Ideally, the instructor should provide the children with some parts of the experiments that require more time to prepare or are more demanding. For example, youngsters will love to make rockets with air propulsion. The instructor should make the platform so the children can concentrate on making different models of rockets using just paper, scissors, and white glue.

The most important point is to create a favorable atmosphere for children to explore new possibilities and to come up with new ideas. Children are very creative, and the instructor is sure to be impressed with their progress and to benefit from many of their suggestions. The success of this strategy relies very much on selecting which activities the kids can do on their own and which ones they need assistance with.

Some projects offer several degrees of sophistication. An example is the wind tunnel, which starts with simply blowing a piece of paper. Another example is the flying saucer ("Outsmarting Friction"), the simplest version of which is a CD model that is easy to make. The next stages of both projects are increasingly more demanding in terms of skills, although they can still be used in demonstrations in connection with the simpler versions. The underlying principles are the same, so both children and adults will marvel at the same experiments. Other projects, although simple to make, are conceptually more sophisticated and are prone to inspire college freshmen. The sequence of experiments of "The Astronaut in the Elevator," "Bouncing Balls," and "New Discoveries with Polarizers" may belong to this class. The experiments which are more demanding are highlighted with a star (intermediate level) and two stars (more advanced level).

One important point to keep in mind is that every age group has different needs and skills. Teenagers 14 to 17 years old will definitely want to do whole projects on their own and put into practice lots of new ideas for science fair projects. They should become familiar with basic tools and safety measures before moving on to implement their projects. This will help prevent accidents. (This holds true for all age groups, for even a simple piece of paper can cut someone's hand if not properly handled—see "A Paper Saw?")

Teachers with no science background but who want to expose their students to the "wonder" of science can greatly benefit from first doing some of the proposed experiments on their own before assigning simple projects to their students. To understand the importance of challenges in the educational process, teachers should first feel for themselves the thrill of discovery. Once they catch the spirit, they will be able not only to enthuse their students but also to explore the results of the experiments and enrich their classes. Parents who home school their children or who want to enrich their children's public school education should also do some of the experiments on their own first, and then do them along with their children. The cooperative effort provides a unique opportunity for parents and children to benefit from their mutual progress as they advance in exploring new possibilities. Students should be encouraged to share their inventions with their classmates and friends so that they can learn from each other as well as with each other.

College and High-School instructors and lecturers who have a laboratory as part of introductory courses should take advantage of the many simple demonstrations proposed here. They should also consider challenging their students to develop innovative projects on their own, using inexpensive materials. The first step toward this goal could be improving projects like the ones proposed in this book, so that students become more motivated, gain confidence, and develop the required skills to tackle more sophisticated projects. I started this approach some years ago and have so far collected a number of success stories. Some of my freshman students were the first ever at my university to apply for a patent.

Since innovation, discovery, perception, and pro-activity are so greatly emphasized in this book, it may also be of interest to non-science students looking for new approaches to the science of everyday life and also for all people involved in fostering creativity in companies, factories, and workplaces in general. After all, we need to move quickly in a world that is technologically reinventing itself every day. Apparently, educational systems around the world are far behind this ever-growing wave of innovative technologies that are shaping a new era for humankind.

WHAT SUPPORT IS PROVIDED

A general introduction to the tools and safety measures needed to implement the projects is presented in "How to Get Going." The material required for each experiment and prototype is described in detail in "Supplies." Suggestions on how to do the projects are presented in "Step By Step." Further suggestions are found in "A Step Further" and "A Step Even Further" sections. Hints about the science underlying the experiments and prototypes, provided in the "Fun Facts" sections, help you find new insights and stimulate further thoughts about the science behind each project. You will be challenged to find out what comes next, to find alternatives when you don't have the materials required, and to invent things you probably never dreamed of. The whole idea is to foster curiosity and fresh explorations. Innovation and discovery are often a product of lack of means, of lucky opportunities that turn up when you are after new possibilities, and of looking at things from different perspectives.

You can also rely on the many sites now found on the World Wide Web. So many of them are dedicated to science and technology that it is impossible to name all. Search engines like Google can lead you to sites full of fascinating information. In my view, though, we should be a bit cautious about the excess of information now available. We live in a world that offers plenty of means but scarcely ever provides us with opportunities to shape our own goals, to fulfill our dearest dreams. If we discover what we really want, what gives us pleasure, what suits our deepest needs, then we can use the incredible resources we have at our disposal in a way that will make us feel proud of achieving something unique and that allows us to give shape to our humanity.

WHY AND HOW THIS BOOK CAME OUT:
THE SPIRIT OF *PHYSICS, FUN, AND BEYOND*

We live in a wonderful world, full of challenges and mysteries, most of them still unnoticed. Physics, like chemistry and biology, represents a human attempt to understand what is behind everything surrounding us, including ourselves. Despite the many advances in science and technology, which are so present in our everyday life, it seems to most of us that only very few people are able to contribute to these advances. As our knowledge increases at a fantastic pace, it seems almost impossible for young people to experience the thrill of innovation and discovery.

The primary aim of *Physics, Fun, and Beyond* is to point out some of the infinite possibilities within our reach to perceive the world from a brand new perspective, using just imagination and the many resources available to us in our homes. Billions of plastic bottles and packaging are produced every year worldwide. Why not use them to invent new gadgets and discover more about the world and ourselves? Do we really need sophisticated equipment to feel the joy of scientific endeavor?

When I was a young student of physics, I came across Richard Feynman's *Lectures on Physics*. Feynman inspired me to take on a big challenge: how to share with everybody the spirit of adventure that makes science and technology advance. This book is a response to a deep-rooted need to bridge the gaps—the many gaps—between advanced researchers and laypeople. Several experiments and prototypes proposed here are linked, one way and another, to current cutting-edge technology and basic research.

The twenty-first century trend is for physics, chemistry, and biology to become more and more intertwined. *Physics, Fun, and Beyond* highlights this trend through experiments and models that display a complementary approach. However, there is more than science and technology in this book. Its most fundamental message is the belief that science and technology are tools to promote people, to help you, me, and other people to discover our talents and improve the overall living standards of our societies. Teamwork, innovation, and challenges are the basic elements that came together to make this book possible.

Many people and circumstances have helped me to transform vague ideas into models and prototypes, or to discard projects that became untenable. It has been a long journey. I started with an attempt to make my physics courses more attractive by challenging the students to create their own projects, and thus become more pro-active. Then I organized interactive exhibitions in shopping malls, parks, squares, and schools, which attracted the interest of people from all walks of life. Workshops with emphasis on creativity and team work then followed. Joining efforts with Alfredo Luis Mateus, who is a chemist and loves art, was a major step toward the creation of the Youth Science Foundation Brazil. It is nice to see that simple ideas and achievable goals can be very powerful in enthusing and inspiring people of all ages to see beauty in science and to apply it to unique creations. This remains my continuing inspiration.

Acknowledgments

I am very deeply indebted to too many people to name them all, people who have taught me how to transform problems and obstacles into opportunities and to persist in my goals. I would like to mention a few representative names: Alfredo Luis Mateus, Esdras Garcia Alves, Andreza Fortini da Silva, Marcus Vinicius de Oliveira Saraiva, Alexsandro Jesus Ferreira de Oliveira, Lister Fleury de Oliveira Laranjo, Sílvio Fernando Vargas Bento, Tiago Novais Faria, Alisson Duarte da Silva, Frederico do Carmo de Moro, Pedro Santiago Carneiro Quadros, Bruna Aparecida de Oliveira, Devair Vieira de Oliveira, Mats Selen, Carlinho, Christian Lanyi and Klaus Weltner. I would also like to express my gratitude to all my colleagues at the Physics Department of the Federal University of Minas Gerais, especially Maurílio Nunes Vieira, Pedro Licinio de Miranda Barbosa, Luiz Orlando Ladeira, Wagner Nunes Rodrigues, Ado Jório, Alaor Chaves, Marcos Pimenta, Oscar Nassif, Luiz Alberto Cury, Ricardo Schwartz Schor, and Beatriz Alvarenga. In addition, Clóvis de Oliveira Mello Júnior, João Batista Reis Silva, and Gilberto dos Santos have been most supportive with their technical advice and prompt assistance. My collaboration with Professors Laurence Eaves and Fred Sheard at Nottingham University in the United Kingdom was also a key element in my search for seeing science from a different perspective.

This book would never have become a reality without the total commitment of Cláudio Roberto, who is responsible for the wonderful figures that are essential to convey the ideas behind all experiments proposed. Heather Jean Blakemore, Michael Hugh Knowles, and Carol J. Lallier also did their best to improve my English. Michael, with his continuous flow of ideas and suggestions, was instrumental in making this book accessible to a wider readership. Bernard Goodwin from Prentice Hall made me believe that *Physics, Fun, and Beyond* was possible. Bernard and Michelle Vincenti did their best to provide me with ideal conditions to write the book. Last but not least, I thank my wife Friederike and our children. Without their understanding, continuous support, and love, I would never have been able to make an old dream come true.

About the Author

Eduardo de Campos Valadares received his doctorate in physics from the Brazilian Center for Physical Research in Rio de Janeiro in 1987 and did postdoctoral research at São Paulo University (1987–1990) and at the University of Nottingham in the United Kingdom (1990–1993). Since 1993, he has been with the Physics Department of the Federal University of Minas Gerais, one of the major Brazilian universities. Valadares has published over 60 papers in different areas of condensed matter physics, physics education, and the popularization of science. His first book is a translation of work by the German poet Stefan George (*Iluminuras*, São Paulo, 2000), followed by *Física mais que divertida* (UFMG University Press, 2000, 2nd edition, 2002), launched in Germany in 2003 by Aulis Verlag Deubner (*Spaß mit Physik*) and now in the United States by Prentice Hall (the American edition, *Physics, Fun, and Beyond*, is enlarged with over 40 new projects and includes comments on all experiments). He also published in Brazil a biography of Isaac Newton (*Odysseus Editora*, 2003), illustrated with low-cost experiments highlighting Newton's ideas. Valadares is co-author of an introductory book on nanotechnology targeted at secondary school teachers, a joint publication of the Brazilian Physical Society and Editora Livraria da Física (2005). In 2001 he received the State Prize Francisco the Assis Magalhães Gomes for his contributions to popularization of science and technology in Brazil. He is also the president of the Youth Science Foundation Brazil. Valadares loves playing with his three children, who deeply inspire him, and to contemplate the world from the top of the mountains surrounding his home.

How to Get Going

The experiments of *Physics, Fun and Beyond* were conceived with safety always at the forefront. Before touching anything, please remember that the simplest and most ordinary materials and tools can cause accidents if they are used incorrectly. Be careful and use common sense when inventing or exploring new ideas.

The majority of the experiments require only tools and materials found at home. It's also a good idea to collect a few additional items in order to build the more complex structures. A brief description of necessary tools follows. Don't skip reading the manuals and safety guidelines that come with the equipment. All the tools listed can be found in a hardware store or other shop that sells tools.

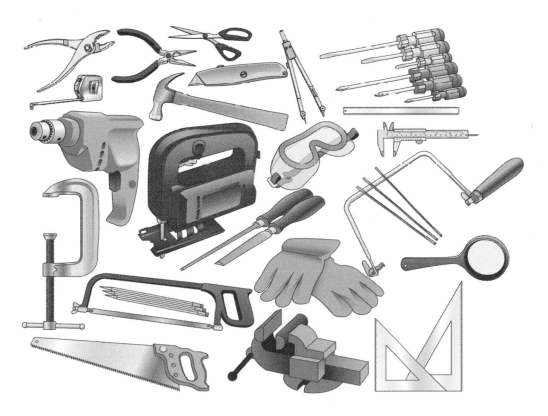

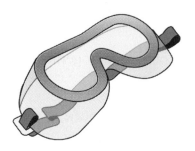

GLOVES AND PROTECTIVE GOGGLES

When you are working with glass, it is essential to wear gloves and protective goggles. Goggles are also very important for any activity that produces dust or small particles in the air, such as sawing or making or enlarging holes in wood or metal.

CALIPER

A caliper is a precision measuring instrument. Calipers made of plastic cost less than metal ones and are more than accurate enough to do the experiments described here. As you can see in the illustration, it is very easy to use this tool.

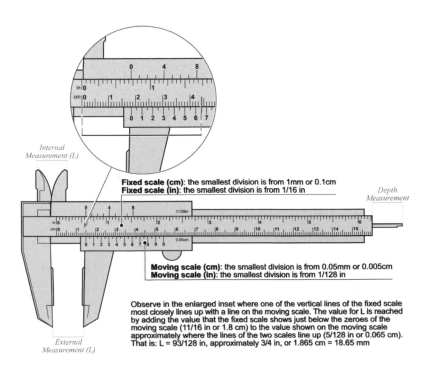

Internal Measurement (L)

Fixed scale (cm): the smallest division is from 1mm or 0.1cm
Fixed scale (in): the smallest division is from 1/16 in

Depth Measurement

Moving scale (cm): the smallest division is from 0.05mm or 0.005cm
Moving scale (in): the smallest division is from 1/128 in

Observe in the enlarged inset where one of the vertical lines of the fixed scale most closely lines up with a line on the moving scale. The value for L is reached by adding the value that the fixed scale shows just below the zeroes of the moving scale (11/16 in or 1.8 cm) to the value shown on the moving scale approximately where the lines of the two scales line up (5/128 in or 0.065 cm). That is: L = 93/128 in, approximately 3/4, or 1.865 cm = 18.65 mm

External Measurement (L)

CLAMPS

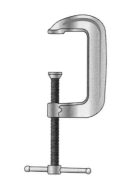

Clamps are used to hold pieces together, for example, when gluing them or to fix them to a workbench. The mid-sized models, between 4 and 6 inches, are well suited to the experiments in this book.

VISE

A vise must be fastened to a workbench. Its basic function is to hold pieces you are working on. It is very useful for tasks like sawing PVC piping or pieces of wood. A small, low-cost model will do for the projects in this book.

ROUND AND FLAT FILES

Both round and flat files are used to file, shape, or sand pieces of various materials. The round file in particular can be used to widen holes in wood, plastic, or metal.

HACKSAW

A hacksaw is a hand-operated saw for cutting wood, metal pipes, and PVC tubes, among other things. Pay attention to the kind of blade you use with it, since different blades are made specifically for use with different materials.

JIGSAW

An electric jigsaw has many purposes, like cutting wood pieces, PVC pipe, and more. Certain blades are made especially for cutting wood and metal. This saw is quite versatile, especially useful for making curved cuts (as in jigsaw puzzles). It can also be adjusted to cut at various angles.

COPING SAW

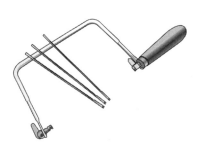

A coping saw is a hand-operated saw for cutting wood, but not metal pipes, and probably not thick PVC pipes either. Its narrow cutting blade (which can break easily if abused) and deep frame (much deeper than a hacksaw frame) allow sharply curved cuts to be made even more easily than with a jigsaw.

ELECTRIC DRILL

The use of a support for the drill makes it easier to use as well as safer and more controllable. To avoid damaging the worktable, the thing you want to drill should be supported by a piece of scrap wood. The choice of drill bits is also very important. There are bits made especially for wood and for metal. Two particularly useful accessories for drills are a sandpaper disk that allows you to sand and finish surfaces quickly and a hole saw for making and enlarging holes. To use a hole saw, you have to first fasten its pilot bit to the electric drill. The pilot bit guides the saw into the wood and keeps it cutting on course. If you are enlarging a hole, there is no wood for

Pilot bit

Hole saws

the pilot bit to bore into. As a result, the hole saw wobbles recklessly all over the place. To overcome this problem, fasten a piece of 1/2-inch plywood over the existing hole. Then bore into the plywood and through the hole behind it. The plywood will hold the pilot bit on track until the saw starts cutting.

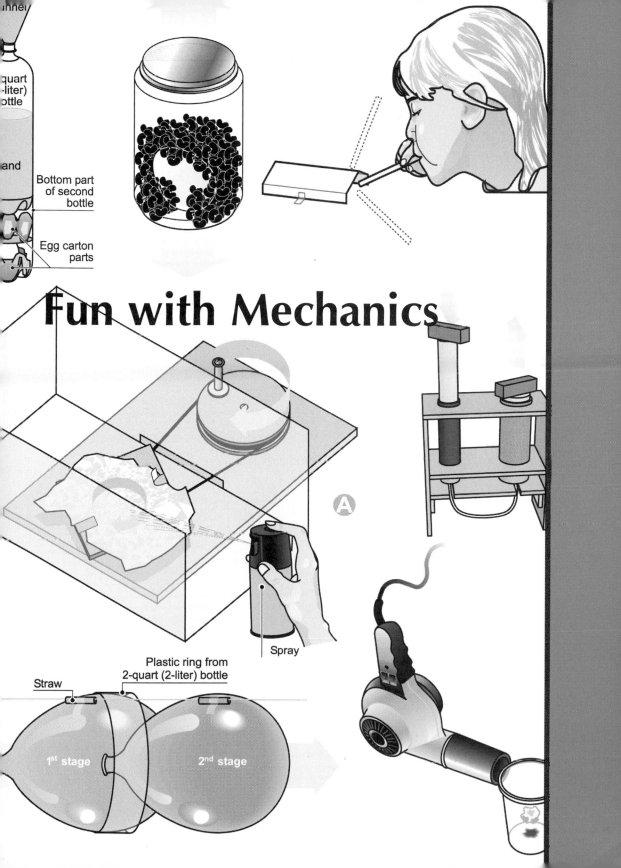

Fun with Mechanics

quart
-liter)
ottle

and

Bottom part
of second
bottle

Egg carton
parts

Ⓐ

Spray

Plastic ring from
2-quart (2-liter) bottle

Straw

1st stage

2nd stage

THE MAGIC CAN

How about making a can that rolls, stops, and obediently comes back to you?

STEP BY STEP

Screw the nut on all the way to the head of the bolt. Make a loop in the middle of the elastic and knot it around the "foot" of the bolt. Make holes in the centers of the lid and bottom of the can. Put one end of the elastic through the hole in the bottom of the can and fasten it with a knot or use one of the nails (see figure). Stretch the elastic and pass the other end through the hole in the lid, fastening it the same way. Shake the can to make sure the bolt doesn't hit the sides. (If it does, you need to tighten the elastic a bit.) Make the can roll on a smooth floor with no obstacles. The magic can will be even more fun if you paint it!

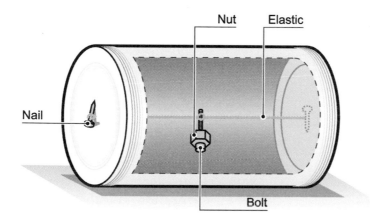

Nut   Elastic

Nail

Bolt

Can you really tame a can? What is the elastic for? What happens to the elastic when you make the can roll? Roll it several times with your hand before letting it go. What happens to the elastic? How about using two elastics instead of just one? How about the bolt? Try a version without a bolt and see if the trick of the magic can still works.

ALTERNATIVE

To see what is going on inside the can, repeat the above, but using a cylindrical, transparent, plastic soda bottle instead of a can. The holes should be made in the exact center of the bottom of the bottle and in the cap. You can also try several other versions with pieces of plastic bottles fitted into each other. Wind-up toys and clocks work in the same way.

FUN FACTS

Stretching an elastic requires some effort. Release it and see what happens. There is something stored in the stretched elastic that can be transformed into motion. This something is called energy. The more the elastic is stretched, the more elastic energy is stored in it. There are other forms of energy. It is not possible to create energy from nothing, only to convert one form of energy into another. Electricity, for example, can make an iron, a TV, a radio, a washing machine, and a computer all function at the same time. All these examples involve the conversion from one type of energy into another. What about the magic can? How is energy stored, and how is it transformed into motion?

HOW THE WEAK BECOME STRONG (STRUCTURING MATERIALS)

All of us have a strong side and a weak side. So do materials. Take advantage of these two sides using rulers, paper, and cardboard.

SUPPLIES

- 2 plastic rulers
- 3 sheets of white paper
- cardboard

STEP BY STEP

A. Bending a Ruler

Hold a ruler by its ends with both hands. Being careful not to break it, try to bend it as indicated in (1) and (2). Can you tell which is the weak and which is the strong side of the ruler? The weak side, seen head on, is wide. The strong side, seen head on, is narrow. How can you take advantage of these two facets of the ruler? You can combine them if you hold up the wider side of one ruler with the narrow side of another ruler, making a T. In this combination, both sides become strong. Other possibilities are cross-sections with the shape of an I (like "I-beams") or a U. These cross-sections are used often in civil engineering constructions. Look in the attic for the beams that hold up the roof boards, especially of buildings made of (or based on) metallic structures.

FUN FACTS

In situation 1 (see illustration), you can think of a ruler that is very narrow in the direction toward/away from the viewer and very thick vertically. In situation 2, the ruler is very wide (toward/away) and very thin (vertically). In both cases, the length of the rulers is the same. Which factor is more decisive for the strength of the ruler in terms of bending? Can you see the advantage of being narrow and thick? The length of the ruler also plays a role in bending. With your hand, hold one end of the ruler on the edge of a table, just like a springboard. By varying the length of the "springboard," discover how it affects the bending of the springboard.

B. Edgy Paper

Try to stand a sheet of paper on its edge. Why does it fall down? Fold the paper, forming an accordion pleat, as shown. Can the paper stand up now? Can it support the weight of a flat sheet of paper? Place the accordion paper between two flat sheets of paper and compare the construction with cardboard. Now make the accordion paper into a bridge, on top of two books, as shown, and find out how much weight it can support. Does your paper bridge support more weight if the number of folds in the accordion plate increases? Now, cut a sheet of paper into two equal halves. Make two identical tubes with each piece of paper. (Use adhesive tape to fasten them.) Make the tubes into a bridge on top of two books, and find out how much weight it supports. Now decrease the radius of the tubes by rolling the paper more tightly. Do the new tubes make a more resistant bridge? How much of the change is due to the decrease in radius and how much to the increased thickness of the tubes?

The mechanical resistance of a body depends on the material the body is made of (compare a sheet of paper and a metal sheet of the same size—see also the next experiment) and on the geometry of the body, as the previous experiments allow you to demonstrate. Folding over sheets of metal can greatly improve their mechanical strength. Different kinds of folding can affect the resistance of a body to bending, stretching, or twisting. (Materials are also structured to absorb shocks.) The sheet metal used in cars, for example, is becoming thinner and thinner. Have you noticed that there are many folds in the bodies of cars? What are they for? Take a look at plastic cups and the bottom of plastic bottles. Look at the various kinds of folds you see in the objects around you and try to discover what they are for. To learn about their properties, use paper to reproduce the folds you observe.

C. More about Paper (Stiffness Test)

From a sheet of paper, cut four 3/4 × 6 in (2 × 15 cm) strips, following the layout illustrated in the figure. Number the strips and the holes left in the paper so that you know where the strips came from. In all combinations, put one strip on top of another and hold them together by one end. Then turn your hand over so that the top strip is on the bottom and the bottom one is on top. Which strip is stiffer? How does the relative stiffness of the strip relate to the direction in which it was cut from the original sheet of paper? What happens to the stiffness of paper when it becomes wet?

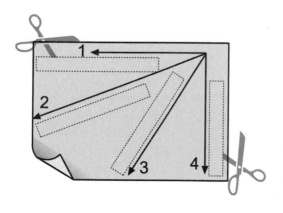

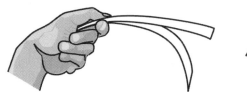

FUN FACTS

If you could look at a sheet of paper through a microscope that enabled you to "see" atoms, you would realize that many of them are all lined up into giant molecules (macro-molecules) called cellulose. These macromole-cules, in turn, form fibers. Ordinary printer paper consists of a network of cellulose fibers linked by hydrogen bonds. You can thus see paper as having a layered network structure, which is almost two-dimensional. (Paper thickness is typically the same as a human hair, around 100 microns, about 10 to 20 fibers placed side by side.) One 8 1/2 × 11 sheet, for example, contains about 10 million fibers. Since cellulose chains also form hydro-gen bonds with water molecules, the fibers swell and soften as cellulose soaks up water. Wet paper is then much less stiff than dry paper because water acts as a plasticizer. The industrial process used to manufacture paper tends to align the cellulose fibers along a cer-tain direction and produce a greater tension of paper along this direction during the drying phase. This results in a greater strength orien-tation. You can also check that by starting a tear in a paper in the direction parallel to the shorter edge of an 8 1/2 × 11 in sheet of paper and watching the tear line gradually curve around toward the perpendicular direction as the paper is torn in half. (Discover also how snags or wrinkles affect the tearing curve.)

Do the stiffness test suggested in C using recycled paper or other materials, like construction paper or plastic. Is there any major difference between industrial paper and recycled paper? Try plant leaves, too!

3

STEPPING ON EGGS

Put eggs to the test and discover how to step on them without breaking them.

SUPPLIES

- egg carton
- two 2-quart (2-liter) plastic soda bottles (cylindrical, with different diameters)
- can lid
- sand
- funnel (paper is okay)

STEP BY STEP

Cut the wider bottle, making a container 8 in (20 cm) high. Cut the egg carton so that you have two circular pieces that can each hold one egg ("upright"). Put one of these egg holders in the bottom of the bottle. Put an egg in it, then the can lid on top of it, as the picture shows, and put the second egg holder on top of the lid, all inside the 2-quart (2-liter) bottle. Then slide the second 2-quart (2-liter) bottle on top of this egg holder and gradually fill it up with the sand. How much weight can the egg hold up? Knowing this, how can you step on eggs without breaking them?

Funnel

2-quart (2-liter) bottle

Sand

Bottom part of second bottle

Can lid

Egg carton parts

A STEP FURTHER

STEP BY STEP

Press the bottle against its bottom, as shown in (A). Now press the bottle sideways, as in (B). In which case is the bottle most deformed? Can you see any similarity between the bottle and an egg?

(A) (B)

FUN FACTS

The neck and the bottom of the bottle are clearly much more "structured" than the bottle sides (the neck and the bottom are curved). When you press the bottle as in (A), you are actually trying to bend the sides of the bottle as in the case of the "strong" side of the ruler that we discovered in Experiment 2, "How the Weak Become Strong." In doing so, you are challenging the strongest parts of the bottle. The situation in (B) is the opposite—like the ruler, the bottle has a weak side. Now replace the bottle with the egg. Can you now see why chicks always hatch by coming out of the eggshells through the sides? By the way, how do you break an egg? And how are eggs positioned in an egg carton? Why can you pile up several egg cartons filled with eggs? Would it be possible if the eggs were placed lying down in egg cartons? The form of an egg alone cannot do its job if chicken feed is not rich in calcium: The material of the eggshell is also very important for its mechanical strength. Without enough calcium, eggshells can become weak and break when transported. In general, engineers have to combine form (geometry) and the strength of the materials they use to make cars, airplanes, rockets, buildings, and many other things.

4   THIN AND FAT BALLOONS

Discover why it is so difficult to blow up a balloon for the first time!

SUPPLIES

- 2 party balloons
- plastic cap with hole at its center or a small piece of PVC pipe

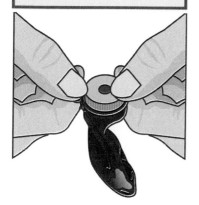

STEP BY STEP

Stretch the lips of a balloon over the cap or piece of pipe. Blow up a full balloon and twist its opening so that the air can't escape, but don't knot it. Stretch the end of this balloon over the cap, so that now the two balloons become connected through the cap. Untwist

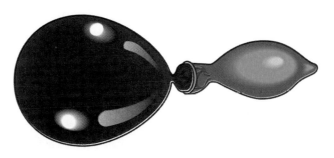

the second balloon opening, allowing the air inside it to expand. What happens to the first balloon? Does it blow up? Press the first balloon. Do the same with the balloon that is more stretched. Which case requires less effort?

A STEP FURTHER

You can imagine a round, filled balloon as a combination of parts of balloons with different radii of curvature, all of them subjected to the same pressure, as in the "Thin and Fat Balloons" experiment. Try to associate the stretchiness of each part of the balloon with its corresponding radius of curvature. Are your findings consistent with the experiment with connected balloons? (Due to the fabrication process, the thickness of a balloon might vary in some places. How does its thickness affect its stretchiness?) The ends of a filled cylindrical balloon can also be seen as two hemispheres with the same radius of curvature as the cylindrical bit. Which parts of the balloon are more stretched: the wall of the cylinder or the hemispheres?

FUN FACTS

With your hands, stretch a piece of a balloon with an object (a coin, for example) placed on top of it. The more you stretch the balloon's piece, the less the object will "sink" (try it!). The more stretched the piece of the balloon is, the more it looks like the wall of the balloon with the bigger radius of curvature. Notice that the object exerts the same pressure on the piece of balloon regardless if it is less or more stretched. The air pressure inside both connected balloons is the same. To stretch the smaller balloon further thus requires a higher pressure. When you start blowing up a balloon, the applied pressure does not create very much tension in its wall for starting the stretching process necessary for inflation. That's why it is difficult to blow up a balloon for the first time. Check it out!

5 PIERCE BALLOONS WITHOUT POPPING THEM

A balloon doesn't always burst when you pierce it. Find its strong and weak sides.

SUPPLIES

- 2 or more party latex balloons
- sewing pin or very sharp needle
- knitting needle
- adhesive tape

Fill up the balloon with air and tie a knot in the neck so the air can't escape. With your pin or needle, carefully pierce the balloon in the part closest to the neck, or on top, as the picture shows. You can also pierce these parts of the balloon as you would pierce an ear, letting the needle come out the other side. Try it with a knitting needle.

A STEP FURTHER

With your pin or needle, pierce the fattest part of the balloon. Of what's left of the balloon after that, pierce a part without stretching it. Take another part and stretch it with both hands. Get someone to carefully pierce this piece.

FUN FACTS

When you blow up a balloon, you have to make some effort. This means that energy is stored in the balloon as elastic energy and in the pressurized air inside the balloon. In the more stretched parts of the balloon, the rubber polymer chains are straightened out and are thus more ordered, like dry spaghetti in a box. By contrast, in the flabby parts, the polymer chains wind themselves up in random coils, like cooked spaghetti. These coils are responsible for the remarkable elasticity of rubber. When you pierce the balloon in a flabby part, the pin or needle is able to enter the rubber without tearing it because the rubber molecules are stretching out of the way of the needle without breaking. (Carefully remove the pin or needle and check whether air squeezes out at the spot you have pierced.) The rubber around the pin or needle becomes distorted and experiences a strong restoring force, filling the gaps and thus preventing the air from escaping. When you pierce a balloon in a stretched part, the balloon tears apart because links and crosslinks between molecules are further broken as the air forces out. In this way, elastic energy is released and the straightened out rubber molecules return to their randomly coiled shape. After all, nature seems to prefer randomness to order.

TWO STEPS FURTHER

A. Anatomy of a Balloon

Examine each part of a filled balloon: its neck, belly, and bottom. Compare round balloons with cylindrical ones. Which parts are flabbier, and which are more stretched?

FUN FACTS

Think of each balloon as a collection of several balloons of different sizes, with the same air pressure inside them. Does the experiment "Thin and Fat Balloons" give you a clue why different parts of the balloon are more or less stretched?

B. Belly of the Balloon

Cover the "belly" of a full balloon with a piece of adhesive tape. Then pierce the taped part with your pin or needle. What happens to the balloon now? Why? Is your balloon now ready for surgery?

FUN FACTS

The belly of a round balloon is the area most stretched of the balloon. (What about cylindrical balloons?). The adhesive tape locally reinforces the balloon so that when you pierce the taped part, the surrounding bits will not feel that the balloon was pierced. The balloon survives with just "minor surgery."

6

STRETCHING CARROUSEL

A spinning object attached to a string makes the string stretch. Build a simple gauge to determine the forces acting on different points of the string.

SUPPLIES

- 1 piece of strong string 2 ft (61 cm) in length
- 2 equal pieces of elastic about 6 1/2 in (16.5 cm) in length
- 2 pieces of black insulating tape about 1 1/2 in (4 cm) each
- 2 lightweight black objects (for example, two small plastic balls)

STEP BY STEP

Securely attach one elastic to each object. Tie with a secure knot the other end of one of the elastics to the far end of the string (opposite the end you will be holding). Then fix a piece of black insulating tape on the string right next to the knot. (The black tape should be on the string and not on the elastic. The black object and black tape will make it easy to see how much the elastic stretches when you are spinning the string.) Securely attach the second elastic to the string with a knot about 1.2 ft (36 cm) from the first elastic. Make sure the length of the elastic from the knot to the object is the same for both objects. Fix another piece of black tape on the string right next to this knot. Now, hold the string close to the second knot and swing it around and around in a vertical plane in front of a well illuminated white wall. Discover which piece of elastic is

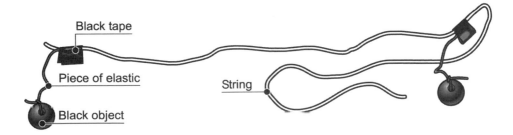

Black tape

Piece of elastic

Black object

String

stretched more as you increase the spinning speed. (Make sure that nobody will be hit by the objects as you spin them or if they fly off the elastics as you spin them.)

FUN FACTS

The objects change direction continuously as you swing the string around and around. To achieve that, a force perpendicular to their velocity is needed. The objects, in their turn, are sluggish and tend to resist any change (they have inertia), so they react by pulling both the string and the elastics. Now, the elastics can be stretched more and more as you increase the force that is pulling them (up to a certain limit!), while the string keeps its length almost fixed. Since the ends of both elastics are fixed to the string, they remain in place. The objects attached to the elastics, on the other hand, resist the changes and thus make the elastics stretch. You can use the stretchiness of the elastics to verify how the force exerted on the objects varies from their distance to your hand, which is more or less the axis of rotation.

7

A PAPER SAW?

If you stretch a piece of paper very tightly, it gets as sharp as a knife. Make paper into a saw in only one step! Here's how.

SUPPLIES

- paper disk, with 3 1/4 in (8 cm) radius
- 2 cardboard disks, 2 in (5 cm) radius
- screw or bolt
- 2 caps from 2-liter plastic bottles
- electric drill
- chalk
- cheap plastic cup
- extra piece of plain paper

STEP BY STEP

With an art/craft knife, carefully cut the top off one of the bottle caps, making a small disk. Drill a hole in the center of the other cap. The small disk and cap will be washers for the screw. Place the paper disk between the two cardboard disks, each perfectly centered, and screw them together into the drill, as the picture shows. Turn on the drill and let the paper touch the chalk. Do the same with the plastic cup. Next, have someone stretch out the extra piece of paper so your "saw" can cut it. Now do you believe in a paper saw? Have you discovered why the paper cuts?

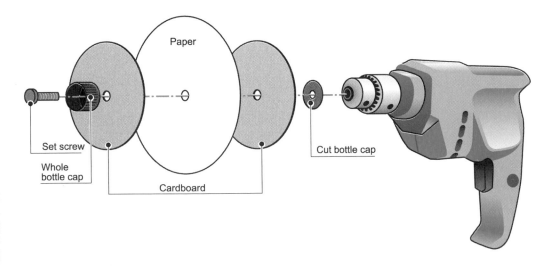

Paper

Set screw

Whole bottle cap

Cardboard

Cut bottle cap

Experiment 6, "Stretching Carrousel," allows you to see what is going on with the spinning disk of paper. Points close to the axis of rotation will feel different forces compared to points at the rim of the disk. Now imagine spinning several identical objects attached to different cords of equal length at the same time, like a carrousel. Does the paper behave in a similar way? Skilled cooks do the same trick when they spin a pizza dough superfast, holding up the dough by just one finger. Can you make sharper saws by changing the radius of the paper disks? What about increasing the speed of rotation of the disk? What would happen if you replace the paper disk with a paper rectangle? Also try paper disks with cuts at different orientations.

---

8 GLOBE OF DEATH

Can you ride a motorcycle upside down? Find out how it can be done risk-free.

SUPPLIES

- party balloon
- marble
- small plastic or Styrofoam ball
- coin
- 2-quart (2-liter) plastic bottle

STEP BY STEP

Stretch the opening of the balloon and put the marble, ball, or coin inside. (Ask someone to help you with this.) Blow up the balloon and hold the opening closed tightly so the air can't escape. Hold the balloon with both your hands and shake it as the pictures show, then stop. What happened to the marble, ball, or coin? Can you make the coin roll on its edge?

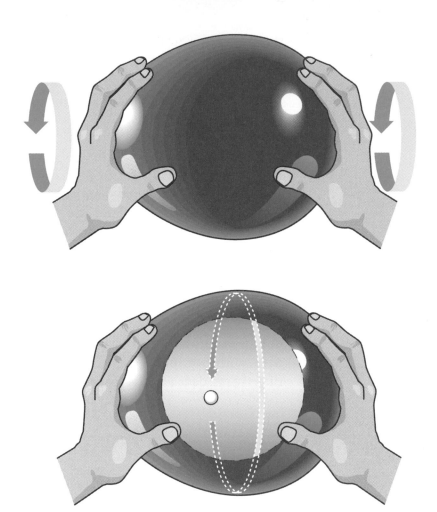

A STEP FURTHER

Repeat the same experiment with the ball or coin, but use a 2-quart (2-liter) plastic bottle instead of a balloon. Keep swirling the bottle around and around, but slowly turn it until the opening is at the bottom. Try to keep the ball or coin from getting out. Then repeat this experiment with a little water in the bottle instead of the ball.

FUN FACTS

This is a simple variant of a spinning object attached to a string. In the present case, the inside wall of the balloon plays the role of the string, since it restricts the motion of the ball (marble). The wall thus exerts a force on the moving ball toward the center of the balloon, which keeps it moving in a circle. The ball, because it has inertia, exerts an opposite force on the wall (it is stretched wherever the ball is, as you can easily see by performing the experiment). So you have action equals reaction. If you increase the spinning speed of the ball, the wall stretches more (which is why you should not spin the ball too fast; otherwise the balloon can burst—see also Experiment 7, "A Paper Saw?"). At its highest point, the ball thus exerts a pressing force on the balloon, pointing upward, opposite the Earth's gravity, and the balloon exerts a downward force on the ball. In response, the ball "falls" down. If the speed of the ball is below a certain limit (try low spinning speeds), gravity takes over and makes the ball fall fast enough that it loses contact with the balloon. Are you now prepared for your next adventure, this time with a roller coaster?

9 FLATTENING THE EARTH AT THE POLES

The Earth was once a ball of hot, soft material that spun around on its axis just as it does today. Let's see how it got flat at the poles.

SUPPLIES

- cardboard
- plywood, 12 × 24 in (30 × 60 cm), 1/2 in (1.3 cm) thick
- piece of solid wire, 10 in (25 cm) long, perhaps cut from a wire clothes hanger
- bolt, extra long, with nut (to serve as handle)
- 1 screw or nail and 4 washers (for axle of larger wheel)
- masking tape
- construction paper
- elastic (sewing) or ordinary string
- plastic tube from a pen
- plastic bottle cap
- superglue
- white glue (for use with paper)
- adhesive tape

STEP BY STEP

Handle

Make a hole in the center of the bottle cap just large enough for the extra long bolt, but small enough that the plastic tube does not fit through it. Cut a piece of plastic tube 1/4 in (6 mm) shorter than the bolt. Pass the bolt through a washer and insert it in the tube. Pass the end of the bolt through another washer and then through the hole in the center of the cap (in from the top). To fasten the handle to the cap, use a washer and tighten the nut on the bolt.

Wheels (One Large and One Small)

To make the larger wheel, cut the cardboard into four disks, two with a radius of 4 in (10 cm) and two with a radius of 3 3/4 in (9.5 cm), marking the center of each. Very carefully, glue the smaller disks together with superglue, making a double-disk. Wrap masking tape around the edge of this double-disk, as indicated. Glue one of the larger disks on top of the double-disk with superglue, using the axle screw or nail to keep the three disks centered. Make a hole in the triple-disk close to its edge, as shown in the figure, to insert (tightly) the bottle cap with the handle attached to it from the bottom of the triple-disk. Glue the remaining larger disk to the bottom of the

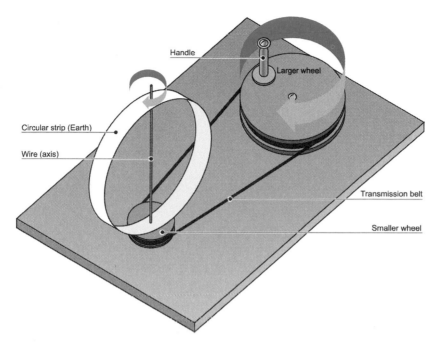

Handle

Larger wheel

Circular strip (Earth)

Wire (axis)

Transmission belt

Smaller wheel

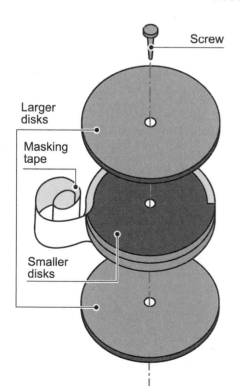

Screw

Larger disks

Masking tape

Smaller disks

triple-disk, keeping all disks centered. Be careful not to glue the wheel to the axle, since it needs to spin freely. To make the smaller wheel, repeat the process with 1 1/4 in (3 cm) and 1 in (2.5 cm) radius disks (no handle is needed here). Make a hole in the center of the wheel for the solid wire to pass through so that the wheel can spin freely.

SET UP THE EXPERIMENT

Cut two strips of construction paper to a width of 1 in (2.5 cm) and length of 24 in (60 cm). Glue one strip on top of the other to make it double-thick. Bend this strip into a circle and glue the ends together to form the "Earth" that will spin. Make two diametrically opposed holes through the strip (at the "North and South Poles" of the Earth). Drill a hole in the wooden board and insert the

Fun with Mechanics

solid wire (axis of rotation) into it, so that the wire is standing up, all the way up to—and through, centered side to side—the top of the circular strip (at the North Pole). Fasten with adhesive tape a small portion of the strip (South Pole) to the center of the smaller wheel. To link the two wheels, use the sewing elastic or ordinary string to make a transmission belt (like a fan belt) by knotting its ends. Place the belt around both wheels, pulling the larger wheel away, across the board. When the larger wheel gets far enough away from the smaller wheel to make the belt stretch tightly, that's the spot where you should fasten the larger wheel to the board with a screw or nail plus a washer at its center.

Watch what happens to the circular strip of construction paper (the Earth) when you turn the larger wheel, producing a movement just like the one the Earth makes as it spins on its axis.

FUN FACTS

Consider two points E and P on the circular strip (the Earth), one close to the Equator (E) and the other close to one of the poles (P). The point E goes around in one large circle in the same amount of time that the point P goes around in a small circle. This means that E travels faster than P. The faster an object

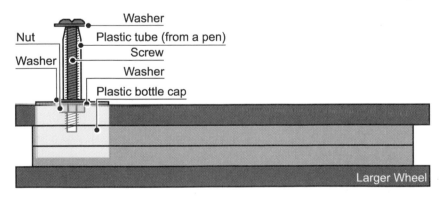

Fun with Mechanics

moves in a circle, the larger the force necessary to hold it in a circle. (See Experiment 6, "Stretching Carrousel," and Experiment 8, "Globe of Death.") Since the strip is not fixed at the North Pole (it can move up and down along the axis), the Earth becomes flat as you make it spin. Try to increase its spinning speed and see what happens. Billions of years ago, the Earth was a giant ball of hot, soft matter, turning around its axis just as it does today. Due to the different forces arising from the spinning of the Earth around its axis, it became flat and remained so as the planet cooled down. What would happen if the Earth was spinning much faster? Would the gravitational forces still hold the different parts of the planet together? Just attach a coin to one end of a thin strip of paper and, holding the other end of the strip, swing the coin around and around, increasing the spinning speed. What happens to the strip of paper? (Make sure that nobody will get hit by a flying coin when you do this experiment.) You can also try to attach one or more coins to the circular strip (the Earth) at different positions using adhesive tape and see how this extra mass affects the form of the Earth as it spins around its axis. (Think about the very massive asteroids and comets that have hit our planet in the past. How might they have affected the Earth's spin?)

A STEP FURTHER: SPINNING BUCKET

STEP BY STEP

Remove the axis of rotation of the Earth and fasten the smaller wheel to the board with the screw or nail plus a washer at its center.

SUPPLIES

- Earth model from the first part of this experiment
- cardboard
- CD case
- Scotch tape (to seal the CD case)
- 1 screw or nail and washer (for axle of small wheel)
- water with food coloring (to contrast)
- superglue

Make sure that the transmission belt is tightly stretched. Cut the cardboard into 4 equal disks (numbered 1 through 4), each with a radius of 3 1/4 in (8 cm), and mark the center of each. Make a hole with a radius of 1 1/4 in (3 cm) in disk 1 and a small hole in disk 2, slightly greater than the washer, so that the smaller wheel can spin freely. Make rectangular holes in the remaining disks (3 and 4) to fasten the sealed CD case ("bucket"), as shown in the figure. Glue the four disks together with superglue, keeping all disks centered, to make a holder for the bucket. Fit the smaller wheel into the circular hole of the compound disk. Fill around a third of the volume of the sealed CD case with colored water and insert the CD case into the rectangular hole. Watch what happens to the water when you turn the larger wheel at different speeds. What would happen if inside the spinning CD case you had two liquids of different densities? The CD case would then operate as a centrifuge.

FUN FACTS

The water's inertia is trying to keep it traveling in a straight line, and the rotating CD case is pulling inward on it to bend its path into a circle. Since the CD case is exerting this inward force on the water, the water must simultaneously exert an outward force on the narrow sides of the rotating CD case (action = reaction).

As Experiment 6, "Stretching Carrousel," demonstrates, the horizontal force acting on the water at different points inside the spinning CD case increases with their distance

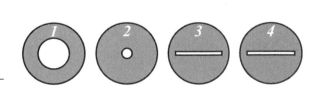

from the center of the CD case (its axis of rotation), where this force is zero. The water is thus pressed against these narrow side walls and at the same time pulled down by the Earth's gravity. The resulting profile of the water surface is a parabola. As you spin the CD case faster, the parabola changes as the inward acceleration of water increases. When the spinning CD case operates as a centrifuge, its contents revolve around the center of the centrifuge (axis of rotation). Inertia tends to make each item go straight, while the CD case makes them bend inward. However, the denser item in the centrifuge has a tendency to travel straighter than the less dense item. The denser items are thus found near the outside of the circular path, and the less dense ones are found near the center of that path. Now, what would happen if the center of the CD case did not coincide with the axis of the small wheel? Check it out for yourself!

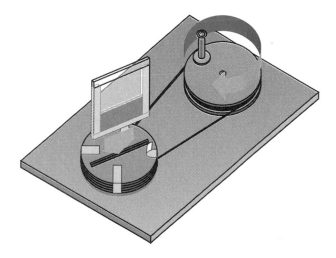

Combine art and science in a fun way by spraying or dropping different colored paints on a rotating canvas. See the "Spinning Bucket" from Experiment 9.

SUPPLIES

- Earth model from Experiment 9, "Flattening the Earth at the Poles"
- round or rectangular piece of cardboard ("easel") with a rectangular hole at its center, as in disks 3 and 4
- piece of paper ("canvas")
- cardboard box
- spray (use, for instance, an empty perfume or deodorant bottle)
- several colors of poster paint
- superglue
- adhesive tape

A. Spraying Paints

STEP BY STEP

Place the cardboard (easel) on top of disk 4 with the holes coincident. Cut two to four strips of cardboard equal in length to the rectangular hole and with a width equivalent to the three layers of cardboard (disks 3 and 4 plus easel). Glue enough of these strips together with superglue so that the resulting strip fits nicely in the three-layer hole. (This will allow you to reuse the project later for other purposes.) The spinning easel, which

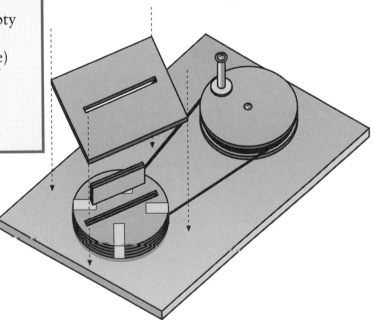

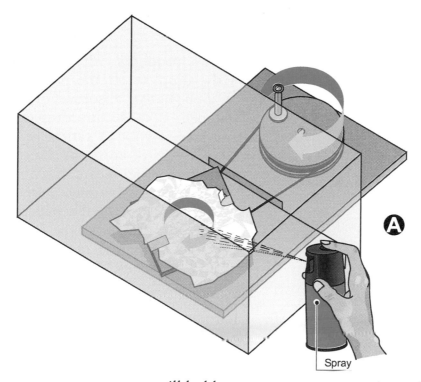

Spray

will hold up your paper canvas, is ready. You can use just four sides of a cardboard box, with a lateral opening as indicated, to make sure that the person turning the larger wheel is not sprayed. Make a hole in the center of the bottom of the cardboard box. Secure the wheel with the screw through the hole in the cardboard box. Now the wheel is inside the box. Make a small slot at the bottom of the box, wide enough for the transmission belt. Scrunch the piece of paper into a ball. Then flatten it out, but not completely. Your canvas should have wrinkles, forming a landscape. Place it on top of the cardboard easel and secure it with adhesive tape. Ask someone to turn the larger wheel while you use the squeeze bottle to spray paint from the side once or twice! Then let the wheel turn the other way and spray another color. Use colors that produce a great contrast (red and green, for instance). You will be proud of your cool painting!

FUN FACTS

The spray produces very tiny droplets. When the droplets hit the paper, they are almost immediately absorbed on the spot. This happens because tiny droplets have a huge surface-to-volume ratio. This means that the number of atoms at the surface of the droplet is a significant fraction of the droplet's atoms. The surface atoms are the ones that interact with the paper surface. The smaller the droplet is, the greater their number, and hence the more efficient the absorption becomes. Because the paper forms a landscape in miniature, like a picture of a chain of mountains with valleys taken by a satellite, if you spray the paint sideways, the droplets will hit only certain areas of the spinning paper. (Which ones?) Try different spinning speeds and use pieces of papers scrunched in different ways to see what happens.

B. Dropping Drops of Colors

STEP BY STEP

Make a hole in the center of the bottom of the cardboard box. This box has five sides, and the open side faces up toward you. Secure the wheel with the screw through the hole in the cardboard box. Now the wheel is inside the box. Make a small slot at the bottom of the box, wide enough for the transmission belt. Place the piece of paper to be painted on top of the cardboard support and secure it with paperclips. While turning the larger wheel, drop drops of paints or inks of various colors onto the rotating paper at different positions (close to the axis of rotation and then away from it). To drop the paints,

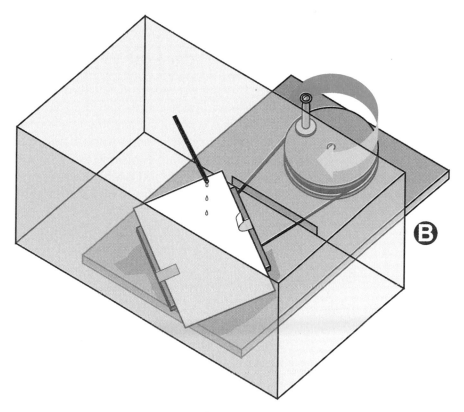

use, for instance, a drinking straw or an eye dropper. Insert the straw into the paint and close the top of the straw. To let a drop fall, just take your finger off the top of the straw for a second, then close it again. Enjoy your masterpiece!

FUN FACTS

Unlike the spray, the drops of paints are big enough to *not* be totally absorbed at the spot where they hit the piece of paper. Therefore, they are forced out as they stand on a spinning platform (see Experiment 6, "Stretching Carrousel"). Depending on where the drops fall and on the spinning speed, you get different results, since the forces acting on them vary with the distance from the spinning axis! Just have a go.

ASTRONAUT IN THE ELEVATOR

You have probably ridden in an elevator and felt that certain floaty feeling when it starts to descend (along with that leap your stomach makes) and pressure in your feet when it starts going up. The following experiments demonstrate what would happen if the cable broke and the elevator suddenly went into free-fall. The same thing happens to a space ship in orbit around the Earth. Actually, it is continuously "falling" in space (known as "free-fall"), like an elevator in a shaft with no end. Check it out.

SUPPLIES

• flexible spring made of wire or plastic

A. Spring Action

STEP BY STEP

Hold one end of a spring and let it stretch downward as far as its own weight will take it. Let it go and see what happens.

B. Water Under Pressure

STEP BY STEP

Make a small hole in the side wall of the bottle about 1 in (2.5 cm) from the bottom. Cover this hole with your finger while you fill the bottle up with water, then close the top tightly. Take your finger off the hole. Now, unscrew the top a little and see what happens. Does the top work like a faucet? Why does the water come out of the hole when you let the air in? Let the bottle fall to the ground (outdoors). What happens to the stream of water?

C. Free-Falling Water

STEP BY STEP

Make a hole in the center of the cap so that the straw just barely passes through it. Seal the straw into the hole with superglue or melted candle wax. With the cap off, fill the bottle so that the straw is partially submerged, as the picture shows. Screw on the cap tightly and blow into the straw until a column of water is formed in the straw. Let the bottle fall (outdoors) and see what happens to this column in free-fall.

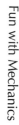

D. Free-Falling Coins

STEP BY STEP

Make a hole in the bottom of the cup, right in the center. Cut the rubber band at one of its ends. With the masking tape, secure the two coins at each cut end of the rubber band. Stick the uncut end—now the center—of the rubber band through the hole in the cup, as shown in the picture. Make a knot there so that the rubber band won't come back through the hole. You can push the match (paper clip or bobby pin) through the knot to ensure this. Then, make sure the band is stretched out to the sides of the cup by the weight of the coins. Now, let the cup fall to the ground. Watch the coins in free-fall.

So, what really happens to the weight of the spring (A), the water (B) and (C), and the coins (D)?

FUN FACTS

Suppose you were in a bus moving on a straight road or on a plane flying along a straight path, both at constant speed. If you cannot see what happens outside, you might think that you were at rest. You can walk along the aisle of the bus or plane and be okay. Now, if the bus or plane suddenly slows down, you will immediately feel it. You will be projected ahead, as if your body wants to ignore the change in speed: hence the importance of seatbelts. The opposite happens if the bus or plane accelerates. In both cases, it seems that your body is accelerated in the opposite direction. This holds also if the bus or plane makes a curve: your body seems to be forced outward. Consider now an object in free-fall. It is continuously accelerated because the Earth attracts it. Suppose you were in an elevator in free-fall. The elevator is accelerated downward, and you are accelerated in the opposite direction, just like in the bus or on the plane. The only difference is that now the Earth attracts you (and the elevator) downward. You are also accelerated both upward (because of the elevator's descent) and downward due to the Earth's gravity). The two effects cancel out exactly, as can be demonstrated through the "Astronaut in the Elevator" experiments. This is called the principle of equivalence (you cannot distinguish between being on a planet or on an accelerated spacecraft in outer space, provided that the acceleration of the spacecraft equals the acceleration in free-fall due to the planet's gravity). The same thing happens to a space station orbiting the Earth. Inside the station, the astronauts feel an artificial gravity, since the station is accelerated as it orbits around the Earth due the Earth's attraction. Both the astronauts and the space station are attracted by the Earth, so the artificial gravity inside the space station and the Earth's gravity cancel out. There is another way to see the absence of gravity in outer space. The astronauts and the space station are continuously falling together, so the astronauts do not press against the inner walls of the space station, and vice versa (like the liquids inside the falling bottles in the preceding experiments). If the space station suddenly accelerates due to its own propulsion, the astronauts will feel a force due to their inertia. They will tend to stay where they are, and the space station will move on, so the astronauts will notice the space station wall ("floor") either approaching them or moving away from them. If the astronauts have no external orientation, they can interpret this as the effect of an artificial gravity. Can you see why astronauts become weightless in a space station in orbit around the Earth?

When you are in a car that brakes or accelerates, it happens parallel to the surface of the Earth. In this case, you do feel you are accelerated: there is nothing to balance the acceleration of the vehicle, as happens in free-fall, since the Earth's attraction only acts downward. Can you now see why, in the preceding experiments, gravity seemed to be switched off?

12 WASHING MACHINE: WATER EXTRACTOR

Take a look at the barrel inside a washing machine. Do you see the whole bunch of little holes it has? What are they for? Why does the barrel turn?

SUPPLIES

- 2-quart (2-liter) plastic soda bottle (cylindrical)
- 1 plastic disk that fits in the bottle (for example, a plastic lid)
- string (strong)
- wet towel

STEP BY STEP

Cut off (and discard) the top of the bottle so that the bottom part (cup) is about 4 in (10 cm) high. Make two small holes about 1 in (2.5 cm) below the top of the large cup on opposite sides. Pass the string through these holes and tie it with lots of slack to make a handle; next, tie this handle to a longer piece of string, as the picture shows. Then, make a lot of small holes in the plastic disk. Put it inside the larger bottle. Put a wet towel or piece of paper on top of the plastic disk. Hold the end of the string and swing the bottle around and around.

What happened with the wet towel? Where did the water go? Replace now the

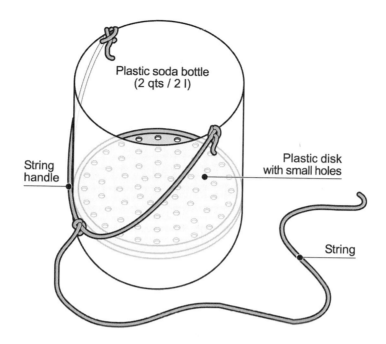

Plastic soda bottle
(2 qts / 2 l)

String
handle

Plastic disk
with small holes

String

plastic disk with holes with the bottom of a smaller plastic bottle (cup) with a single hole. Where should the hole be if the "barrel" now spins around a vertical axis (twist the string and let the "barrel" go)? (Hint: Put some water in the cup to test this new variant.) So, now do you know why the barrel of the *washing machine* has all those little holes?

A STEP FURTHER: WATER UPSIDE DOWN

STEP BY STEP

SUPPLIES

- model from the first part of this experiment
- water

Fill the cup with some water. Hold the end of the string and swing the cup with water around and around in a vertical plane. What happens to the string? What about the water? Does it fall down to the ground?

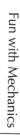

FUN FACTS

As you make the cup spin over your head, you are pulling downward on it and thus making it accelerate downward very rapidly. The water and wet towel remain in the inverted cup because the cup is accelerated faster than gravity alone can accelerate both. Although the water and the towel are free to fall, the cup overtakes the falling water and towel. As a result, the water and wet towel are pressed against the bottom of the cup and plastic disk. Because only the water can pass through the small holes in the disk, you separate the water from the towel. There is another way to see it. When the cup is accelerated, as you swing it around, an artificial gravity is created inside it, opposed to the cup's acceleration, which is along the stretched string. This artificial gravity is greater than the Earth's gravity and thus prevents the spinning water and towel from falling when the cup is upside down. The water and wet towel are then pressed against the bottom of the cup, and the water separates from the towel as it passes through the holes.

13   THE SQUARE WHEEL AND OTHERS

Why not have a square wheel? Take advantage of this unique opportunity to reinvent the wheel and the roads.

SUPPLIES

- 2 pieces of wood measuring 3 3/4 × 52 in (9.5 × 132 cm)*
- 4 pieces of wood measuring 2 3/4 × 5 1/8 in (7 × 13 cm)*
- wooden board to cut the wheels from
- wooden broom handle (a dowel will also work)
- wood glue
- patterns found on pages 333–334

STEP BY STEP

The Speed-Bump Road

Align the two longer pieces of wood one on top of the other so that the sides and ends match precisely, and (temporarily) nail them together, keeping in mind that the nails must keep the strips firmly together and aligned through the whole process of cutting the double thick strip to the road pattern. (For safety, this should be done on a workbench, using short nails and with a scrap piece of wood underneath.) Keeping the strips together and aligned ensures that the two rails of your "speed-bump road" will be identical. Copy the road pattern on pages 333–334 at least five times, drawing one after the other on the top wooden strip, keeping the lower edge of the pattern even with the lower edge of the wooden strip (see the illustration). Cut the double-thick strip according to the pattern you've drawn on them. Use a coping saw or a jigsaw and *adult supervision*. Then take out the nails. To keep the rails parallel, nail the four rectangular pieces (5 1/8 in/13 cm) in between the two rails.

* All of the wooden pieces should have the same thickness of 3/4 to 1 in (2 to 2.5 cm).

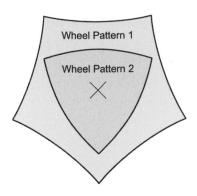

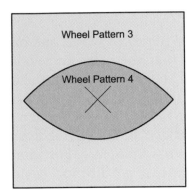

The Wheels

Use the wheel patterns of your choice (see pages 333–334). Make sure the two pieces you cut from the wooden board are big enough for the pattern to fit inside. Nail the two pieces together, as you did for the rails (two nails should be enough). Trace the pattern on top, then cut out the wheels. With a drill, make a hole right in the center of each X printed on the patterns provided. This hole should be just a little smaller than the broom handle (or dowel) so that it can fit tightly with no slippage. Start with a somewhat smaller hole and make it larger until you get the right size. Then, use a file to carefully smooth out the sides of the hole without making it too big. Take out the nails holding the wheels together. Saw the broom handle into various pieces of 6 3/4 to 7 1/8 in (17 to 18 cm) in length. Keeping the points of the wheels carefully aligned, put glue on each end of the broom handle pieces and insert them into the holes. These are now your axles. As a final touch, you can sand and paint your wheels and road. Have fun with your new invention.

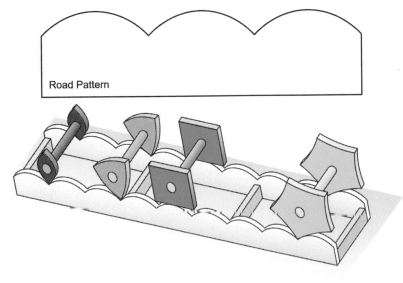

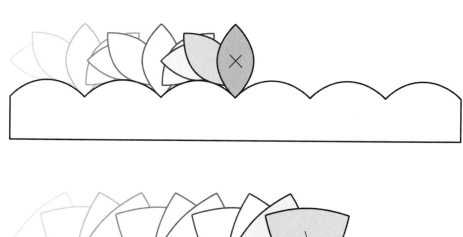

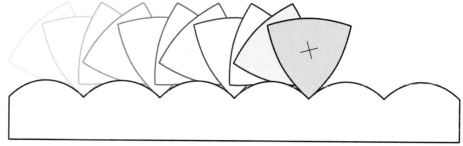

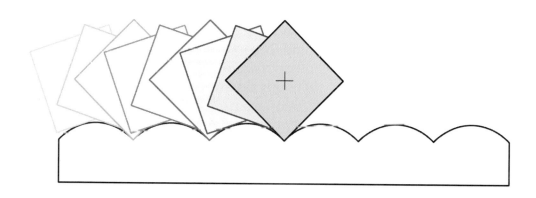

On the speed-bump road that you just made, the only wheels that can roll smoothly are of certain shapes and sizes, like those in the patterns provided. Have a close look at the road profile and at the wheels to see how they match each other, thus keeping the axis of each wheel always at the same level as the wheels roll. Are the center of the wheels (axis) and their contact points vertically aligned? The smallest wheel is a polygon with two curved sides. The next largest wheel could be seen as a curved triangle, followed by a square and a pentagon. We can treat other wheels as polygons with varying numbers of "curved sides": 6, 7, 8, 9. As the number of sides of the polygons increases, so does the size of the wheels. With an ever-larger number of sides, the wheel starts to more closely resemble a round wheel. In this case, the bumps in the road become insignificant compared to the relative size of the wheels. We find something like this situation in the world of atoms.

In atoms, electrons have only certain energies, unlike in our ordinary world, where people, bicycles, cars, and airplanes can have energies that vary continuously (a car or a plane can be accelerated continuously). You can go up a ramp, but an electron in an atom can only go up "stairs." On a flat road (our daily life experience), wheels of *practically* any size radius can roll evenly, unlike they can on the speed-bump road (the world of atoms). A giant, circular wheel won't be able to "feel" the tiny bumps of the atomic speed-bump road, much like the real-life case of a tire rolling on sand. For the same reason, we cannot perceive the tiny atoms in our world of giants. Did you know an atom is to you in size roughly as a flea is to the whole Earth?

14

BALLOON ROCKETS

It's easy to turn a balloon into an air-powered rocket, with one or two stages. Amaze your friends!

SUPPLIES

- balloon
- nylon fishing line or string, 6 to 8 yd (5 to 7 m) in length
- drinking straw
- masking tape

A. One-Stage Rocket

STEP BY STEP

Use the nylon line as a guide. Tie one of the ends of the line at a high point. Cut a piece of the straw 2 to 4 in (5 to 10 cm) long and thread the line through it. Blow up the balloon and hold its opening closed tightly with your fingers so no air escapes. Ask someone to tape the straw piece to the balloon, as the picture shows. Now, stretch out the line and let your rocket go!

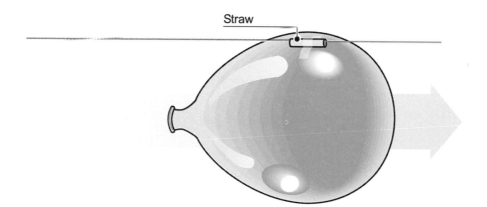

Straw

B. Two-Stage Rocket

STEP BY STEP

Cut two pieces of the straw to lengths of 2 to 4 in (5 to 10 cm), and run the line through them. Cut a ring out of the bottle with a width of 3/4 to 1 1/4 in (2 to 3 cm). Use the same line from the previous experiment. Fill one of the two balloons (the second stage of the rocket) and hold its opening closed with your fingers. Stretch the opening and press it against the wall of the plastic ring. Get someone to blow up another balloon (first stage of the rocket) so that the widest part of this balloon is inside the plastic ring. When it's full, the opening of the second balloon will be trapped against the wall of the ring and will not lose its air, as the picture shows. Hold the first-stage rocket balloon tightly closed with your fingers. Get someone to tape the straw bits to the balloon with the masking tape. To see your two-stage rocket in action, stretch the line and let go of the neck of the first-stage balloon.

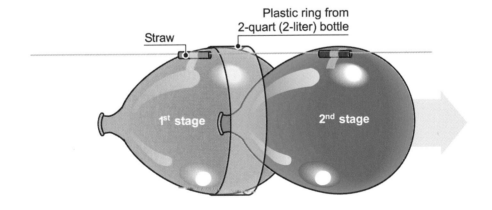

Straw

Plastic ring from
2-quart (2-liter) bottle

1st stage   2nd stage

When you blow on a large, light object (like a filled party balloon), it moves easily. The air stream hits the object and the object hits the air back. When you jump, you do the same: you push the Earth and the Earth pushes you back (action = reaction). At the balloon's open mouth, pressure energy and elastic energy are converted into kinetic energy. So, the air coming out of the balloon moves to the left (see figures). The balloon actually kicks the air out. The air leaving the balloon, due to its sluggishness (inertia), would prefer to remain stationary. It thus reacts by kicking the balloon back, so the balloon moves to the right. The balloon's mass decreases continuously as the air comes out of it, just as a real rocket loses mass as more fuel is ignited, producing jets of gas that are thrown away at the bottom of the rocket. You can use the same trick to change the rocket's trajectory laterally: you just have to throw away gases from an opening in the rocket's lateral wall. This will force the rocket to move in the opposite direction (it is again the action = reaction principle at work).

15

ROCKETS WITH CHEMICAL AND AIR PROPULSION

Let's build a cool rocket and its launch platform from scratch.

SUPPLIES

- vinegar
- baking soda
- empty film canister with lid
- construction paper
- superglue
- masking tape

A. Chemical Propulsion

STEP BY STEP

Put a small amount of baking soda and some vinegar (figure out for yourself the proportions that work best) in the film canister and cover it quickly. The lid of the film canister is the bottom end of your rocket. Stand back and get ready for blast-off! You can improve your rocket by adding a conic nose and fins made of construction paper to it, as done in the next model (air propulsion).

WARNING

Be sure you choose a launch site that will not be damaged by the vinegar and baking soda!

A STEP FURTHER

There is a slightly more sophisticated way to engineer your rocket. Mix the baking soda with water to form a paste that will stick inside the lid of the film canister even when you hold it upside down and put it on the canister. Calculate the right amounts, spread the baking soda inside the lid, and place the vinegar in the canister. Put the lid on the canister, keeping the canister upright. When you are ready for launch, turn the canister upside down and place it on the launch site. Watch it go!

Can you think of other ways to engineer a vinegar and baking soda rocket?

FUN FACTS

In the case of the chemical propulsion, vinegar, an acid, and baking soda (sodium bicarbonate) mixed together produce carbonic acid, which quickly decomposes into carbon dioxide (CO_2) and water. As the CO_2 expands, it makes the rocket (canister) move up. The rocket is hit by the CO_2 molecules moving up and, in turn, it kicks them back. It is the action = reaction principle at work.

B. Air Propulsion ★

STEP BY STEP

The Platform

With a hacksaw (or other saw), cut off a piece of the wood stick. Attach the stick laterally to the base with two nails, as shown in the figure. Insert one end of the hose into the tube until it reaches the tube's end. Drill a hole all the way through from one side of the tube to the other to pass the screw. Use the screw to fasten the tube to the wood stick. Insert the other end of the hose into the bottle (if necessary, use masking tape around the tube's ends to keep the hose tightly fitted).

The Rocket

Make an 8 in (20 cm) long cylinder out of construction paper, which fits tightly in the PVC pipe. Draw on the construction paper a 1 1/4 to 2 in (3 to 5 cm) radius semicircle,

---

SUPPLIES

- construction paper
- 3.3 ft (1 m) piece of ridged hose
- 9 in (22 cm) piece of PVC pipe in which the hose can be fitted tightly
- 2-quart (2-liter) plastic bottle (you can also use 1-gallon bottle)
- piece of wood (base), roughly 4 × 10 in (10 × 25 cm), 5/8 in (1.5 cm) thick
- wood stick, roughly 1 × 1 × 8 in (2.54 × 2.54 × 20 cm)
- 1 nail
- 1 screw with washer
- piece of wire
- disk, 1 1/4 to 2 in (3 to 5 cm) radius, or a drawing compass
- masking tape
- white glue
- small Styrofoam ball (optional)

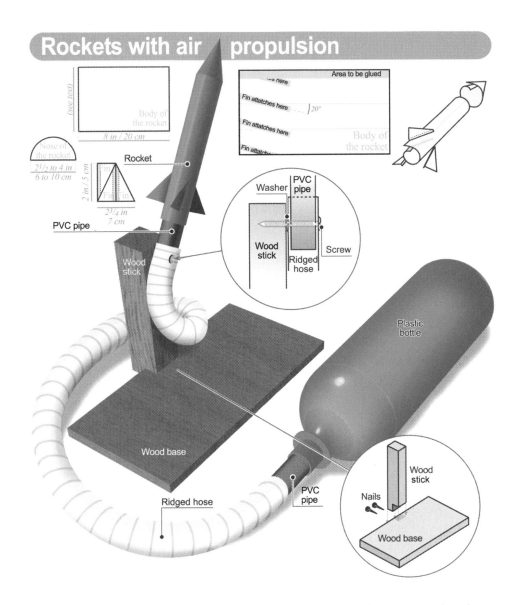

Body of the rocket

(see text)

8 in / 20 cm

Nose of the rocket

2¹/₂ to 4 in
6 to 10 cm

2 in / 5 cm

2³/₄ in
7 cm

Rocket

PVC pipe

Wood stick

Area to be glued

Fin attatches here

Fin attatches here

Fin attatches here

Fin attatch

20°

Body of the rocket

Washer

PVC pipe

Wood stick

Ridged hose

Screw

Plastic bottle

Wood base

Ridged hose

PVC pipe

Nails

Wood stick

Wood base

using the disk as a guide, or use the drawing compass. Cut out the semicircle and make a cone out of it that fits the cylinder. Attach the cone to the top of the cylinder with masking tape. Alternatively, you can use a Styrofoam ball as the nose of the rocket. Just fit it in the top end of the cylinder (using glue or masking tape to fasten it to the body of the rocket). Cut out three rectangular triangles with roughly 1 3/8 in (3.5 cm) base and 2 in (5 cm) height for the three fins. Fold 1/4 in (0.5 cm) of the border of the triangles (see figure) and attach them to the rocket's body at the places indicated. The fins can be positioned either vertically or slightly tilted (around 20°). To launch the rocket, just put the bottle on the ground and stomp on it with one foot. You can try different rocket designs to see which achieves the highest point. You can also tilt the platform and discover at which angle the rocket reaches the maximum distance on the ground. After each launching, the plastic bottle deforms. To restore it, you have to blow through the top end of the PVC pipe. Eventually, you will have to replace the bottle and make a new rocket.

FUN FACTS

The air flow kicks the rocket up, and the rocket in its turn kicks the air back. This is just the action = reaction principle at work. Can you recognize here the same principle of the previous rockets? The fins of the rocket also play an important role, as they do for sharks and other fish. When the fins are tilted, they make the rocket spin around its axis. When you ride a bicycle, you don't fall because the spinning wheels give it a special stability. (See Experiment 35, "The Bicycle's Trick") The same holds for the turbines of the jet engines of an airplane. What about the rocket?

Discover how simple plastic bottles can become rockets that can shoot up to 40 feet (12 meters) high.

SUPPLIES

- two 2-quart (2-liter) plastic bottles
- 2 screw-on caps (the same as came with the bottles)
- bicycle pump
- 5 ft (1.5 m) length of hose that can attach to the bike pump without air escaping
- masking tape
- water

STEP BY STEP

Cut the top off of one of the plastic bottles so that it will stand 6 in (15 cm) high: this is the launch platform. Cut an opening in the side of this bottle, close to its bottom, for the hose to come out. Cut another opening in the center of the cap of the second bottle (the rocket) just large enough for the hose to go through, with no slippage (except at the desired high pressure). Put about 1/2 pint (250 ml) of water in the second bottle (rocket). Screw the cap on tightly and fit the hose through the hole in the cap. Place the rocket on its launch platform, upside down, as the illustration shows. The hose should pass through the hole you made for it in the first bottle. Fit the other end of the hose onto the bicycle pump. To launch, simply start pumping the air in.

HELPFUL HINT

First try launching the rocket without the water, simply by pumping in the air, and see what happens. Then, try it with the water and observe the difference. What's so special about that water? Try using more or less water and find out how it affects the maximum altitude your rocket can reach. Also try this with smaller rockets, using 1-pint (1/2-liter) bottles.

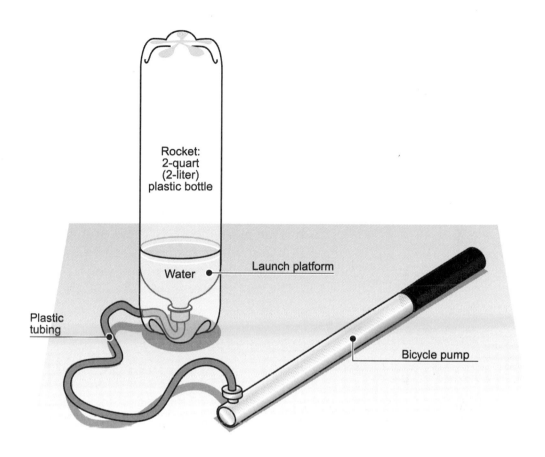

Rocket:
2-quart
(2-liter)
plastic bottle

Water

Launch platform

Plastic
tubing

Bicycle pump

A STEP FURTHER

You can try to improve the performance of
your water rocket using simple accessories.
Cut off a 1-gallon plastic bottle with screw-
on cap to obtain a funnel (nose of the rocket). Attach the funnel (with cap screwed on
tightly) to the top of your rocket using masking tape. Make three fins out of cardboard
and attach them to your rocket, as done in
the rocket with air propulsion. Compare the
performance of your rocket with and without
these accessories. You can try each one separately (fins or nose), and then together.

Fun with Mechanics

This is a variation of the balloon rockets. You just need to replace air with water. As you pump air into the rocket containing water, the pressure inside the rocket increases until it becomes high enough to expel the water. As the water flows out, the rocket becomes lighter and moves in the opposite direction of the water (just like the air in the balloon rockets). Can you see the action = reaction principle at work here? Is there any advantage in using water instead of air? Can you combine the two ways of propulsion?

17 BOUNCING BALLS

Hold the large ball in one hand and the small ball in the other hand. Let them fall from the same height at the same time. How high does each one bounce up from the ground? Now let's try something different.

SUPPLIES

• 2 rubber balls: one large and one small

STEP BY STEP

Hold the small ball on top of the large one at a height of approximately 3 to 4 ft (1 meter). Let the two balls go.

What happens to the small ball when the large one kicks back after hitting the ground? Can you see how a pedestrian (small ball) can be thrown so far away when hit by a car (large ball)?

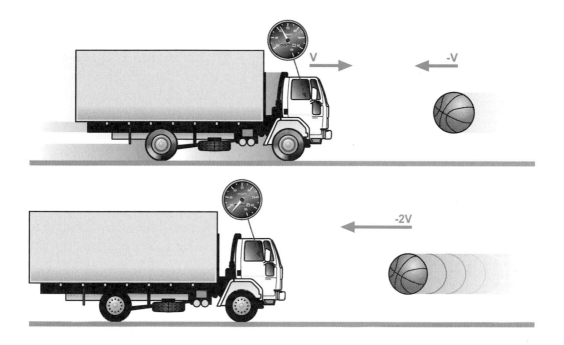

A STEP FURTHER

Repeat the experiment using three balls of different sizes. Put the medium-sized ball between the big and the small ones. Compare how far the small ball bounces in each of these two experiments.

FUN FACTS

Imagine a ball hitting a parked truck head on. The ball will be kicked back with almost the same speed it had just before the collision, while the truck will stay right in the same place. This happens because the ball is much lighter than the truck. Suppose now that the ball and the truck are moving toward each other at the same speed ($v_{ball} = - v_{truck} = v$), as detected by stationary radar. Imagine that you are sitting on the truck, watching the collision. For you, it is as if the truck were parked ($v_{truck} = 0$) and the ball is approaching you at twice the speed detected by the radar ($v_{ball} = 2v$). After the collision, the ball will have the same speed ($2v$), except that now it moves in the opposite direction. To be consistent with the stationary radar, you have to add the speed of the truck to the speed of the ball. (For the radar, both ball and truck are now moving in the same direction.) As a result, the final speed of the ball according to the radar will be $3v$. Can you see the connection between the truck and ball collision and the two bouncing balls?

In the case of the three bouncing balls, the above reasoning indicates that the intermediary ball, after colliding with the largest ball, will move upward with a speed approximately equal to $3v$. It will then collide head on with the smallest ball, which is moving downward with a speed equal to v. Can the use of an intermediary ball really further increase the maximum height reached by the small ball?

18

TEMPERAMENTAL PENDULUMS

Did you know that pendulums are temperamental? How about radios? And TV? When they are provoked, each one reacts in its own way. Check it out.

SUPPLIES

- wooden broom handle
- thick wire
- 2 pieces of 3/4-inch PVC pipe, 4 in (10 cm) long
- 4 identical objects, such as wooden blocks or lead fishing weights (sinkers)
- 3 feet (90 cm) of nylon fishing line
- 2 pieces of wood, roughly 1 × 8 in (2.5 × 20 cm) with a thickness of 5/8 to 1 in (1.5 to 2.5 cm)
- wooden base, roughly 2 × 16 in (5 × 40 cm) with a thickness of 5/8 to 1 in (1.5 to 2.5 cm)
- screws or nails

A. Rhythm and Swing

STEP BY STEP

Build the base with the 2 × 16 in (5× 40 cm) wooden base and the two 1 × 8 in (2.5 × 20 cm) pieces, making an "I" (see illustration). Cut two poles of 1 ft (30.5 cm) each from the broom handle. Screw or nail the poles upright with a distance around 1.2 ft (36 cm) between them. Make a hole in one of the PVC pipe pieces 1 in (2.5 cm) from the top. In the second piece of PVC pipe, make two holes by entering one side and coming out the other at the same height, so the wire can pass straight through both. Bend the wire so that you have five small dips, spaced 2 3/4 in (7 cm) apart, the first one 3 1/8 in (8 cm) from one end. Put that end of the wire through the one hole in the first pipe and the other end through both holes in the second pipe. Then fit the pipes on top of the wooden poles. Note that the last dip in the wire (the fifth and rightmost in the figure) is to keep it in place. Cut four pieces of the nylon fishing line into lengths of 0.4, 0.6, 0.8, and 1 ft (12, 18, 24, and 30.5 cm). You can also try other lengths. In order of shortest to longest, tie the ends of the wires onto the four dips (excluding the last one, very close to the pole). Then tie the four identical weights (blocks of wood or fishing lures) onto the hanging ends of the

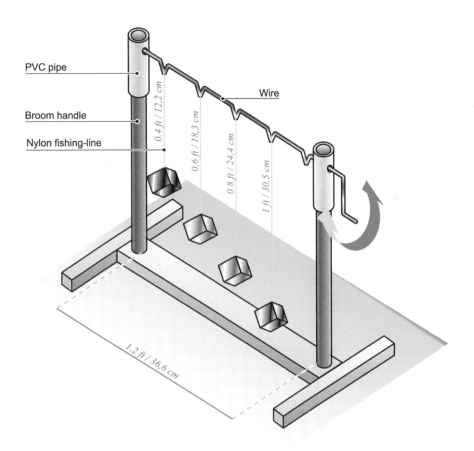

PVC pipe

Broom handle

Nylon fishing-line

Wire

0.4 ft / 12.2 cm

0.6 ft / 18.3 cm

0.8 ft / 24.4 cm

1 ft / 30.5 cm

1.2 ft / 36.6 cm

lines (also try different weights and see what happens). Bend the wire that extends out of the pole into a handle, as the picture shows.

Swinging Mode

First, pull aside two neighboring weights and release them at the same time to compare their rhythms. Is their frequency (number of complete turns per second) the same? Now, swing the handle back and forth with a steady rhythm. Little by little, swing it a bit faster. Compare the swinging mode of the handle with that of each pendulum. What happens when the frequency at which you swing the handle coincides with the frequency of one of the pendulums?

- broom handle or PVC pipe
- 2 soda cans, open and empty but with their pop-tops still attached
- string
- water

B. The Can-Can

STEP BY STEP

Cut two pieces of string to lengths of 0.8 and 2 ft (24.4 and 61 cm). Tie one end of each string onto each of the metal pop-tops. Tie the other ends of the strings 2 ft (60 cm) apart onto the broom handle so that they hang down, as the picture shows. Fill the cans halfway with water and hold the broom handle high. Swing the handle in a slow and steady rhythm, and then try it at higher frequencies. Which pendulum reacts more quickly to the swinging (at a given frequency)?

You can also find out if two pendulums with the same length string but different masses (different volumes of water or sand inside the cans) have different swinging frequencies. Does the kind of material inside the cans affect their swinging? (Compare two pendulums with equal lengths but different material inside the cans, sand and rice, for instance.)

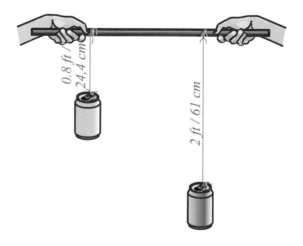

0.8 ft / 24.4 cm

2 ft / 61 cm

FUN FACTS

When you swing the handle, you transfer energy to the pendulums. By performing the experiment, you can find out when this energy transfer is most efficient. In this case, we have what is called *resonance*. You can think of it in terms of a conveyor belt in a factory used to transfer goods. Suppose that the goods are regularly loaded on the conveyor at a certain point and collected at another point. If the loading frequency equals the collecting frequency, all loaded goods will be successfully collected and piled up. In this analogy, the goods represent the energy associated with the swinging of the handle, and the piled goods represent the amount of energy transferred to a pendulum at resonance (the extreme positions of the weight become increasingly farther away from the vertical position of the pendulum). You can use the pendulums model to explain how it is possible to tune into a specific radio (AM or FM) or TV station. Different stations transmit continuous signals in different frequency bands. When you tune your radio or TV, you set the specific frequency (band) you want to receive. Your receiver becomes sensitive only to that single station at the frequency you selected, the resonance frequency. With the pendulums, it is as if you had various receivers tuned in to different frequencies or stations (the pendulums) with only one of them clearly receiving your transmission (the back and forth movement of the wire). Check it out.

19

HYPERSENSTIVE RINGS

See how paper rings react when you shake them up.

SUPPLIES

- 2 sheets of white paper
- cardboard
- adhesive tape
- ruler
- scissors

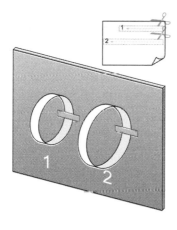

STEP BY STEP

A. Ring Size

Cut two strips of paper out of one of the pieces of paper. Cut them in the same direction, each with a width of 3/4 to 1 in (2 to 2.5 cm) and with lengths of 0.6 and 1 ft (19 cm and 30.5 cm). Experiment later with other lengths! Make the ends of the strips meet so that the strips form rings, and tape each of them (where the ends join) onto the cardboard, as the picture shows. You are ready for action. Shake the cardboard, first side to side and then up and down. Each time, start shaking the cardboard in a slow rhythm (low frequencies) and then increase your speed gradually until the two rings are "dancing." What can you say about the size of the rings and the frequency each one is sensitive to? What kind of connection can you see between this experiment and the previous one, "Temperamental Pendulums"?

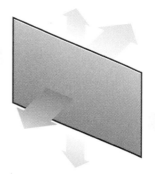

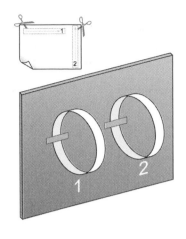

B. Ring Substance

Cut two strips of paper of the same size, but in perpendicular directions, as shown. They should both be 3/4 to 1 in (2 to 2.5 cm) wide and 8 in (20 cm) long. Number each strip and its corresponding direction on the sheet of paper and do the resistance test from Experiment 2, "How the Weak Become Strong." Then join the ends of the strips to form rings and tape them (where the ends join) onto the cardboard, as you did in part A of this experiment. Shake the cardboard as you did before, following the same procedures. How does the stiffness of the paper relate to the frequency that each ring is most sensitive to?

C. Twisted Ring

Cut two strips of paper of the same size and in the same direction (parallel). They should both be roughly 3/4 to 1 in (2 to 2.5 cm) wide and 8 in (20 cm) long. For one of them, join the ends to form a ring, just as in the previous experiments, taping it (where the ends join) onto the cardboard. For the other one, twist one end 180° and join the two ends, as the illustration shows, making a "Möbius strip." Then, tape it (where the ends join) onto the cardboard, a good distance from the other ring. They should look like the illustration. Again, shake the cardboard, slowly at first then gradually faster, as in parts A and B of this experiment. How does the form of the strips affect the frequency each one is sensitive to?

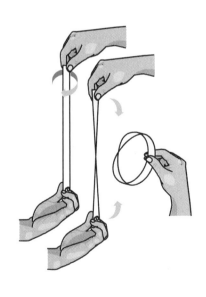

Notice that the Möbius strip has a bend, while the other doesn't. Remembering Experiment 2, "How the Weak Become Strong," what advantage can you say that bending the paper gives it? Have you already

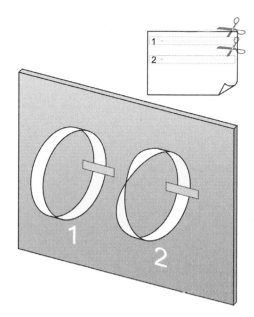

thought to try mixing experiments B and C by making Möbius strips of first one and then the other perpendicularly cut strips?

A STEP FURTHER

Do the same test with other materials, like construction paper or plastic. As Experiment "Slow Motion Camera" shows, you can use a computer monitor screen or TV to see the oscillations of the rings more easily.

FUN FACTS

When you shake the cardboard, you make the paper rings dance. Each ring has its own dancing rhythm with a corresponding frequency (number of "performances" per second). When the shaking frequency matches a ring's frequency, you have a resonance. In this case, the transfer of energy from the shaking to the ring is optimal. Engineers are very keen on determining the resonance frequency of structures (like bridges, skyscrapers, and airplane wings) to avoid resonance effects that could damage them. Buildings of different heights, for instance, are affected in different ways during an earthquake, depending on the frequencies that the earthquake puts the most energy into. The idea is to protect structures by using special materials and geometries so that they are not adversely affected by external oscillating disturbances, such as wind, earthquakes, and so on.

Fun with Mechanics

Make your own bed of nails and discover the secrets of the Indian fakirs!

SUPPLIES

- wooden board, 1 × 1 ft (30.5 × 30.5 cm) and 1/2 to 3/4 in (1.5 to 2 cm) thick
- 2 wooden strips, 1 5/8 in × 1 ft (4 cm × 30.5 cm) and 1/2 to 3/4 in (1.5 to 2 cm) thick (frame edges)
- 2 wooden strips, 1 5/8 in × 0.73 ft (4 × 22.5 cm) and 1/2 to 3/4 in (1.5 to 2 cm) thick (frame edges)
- 400 nails, roughly 1 1/8 in (3 cm) length
- 12 nails to nail the frame together
- balloons (a few)

STEP BY STEP

CAUTION ——————————————

Adult supervision is advised.

Draw a square 8 × 8 in (20 × 20 cm) in the center of the board. Now, draw a regular grid pattern within the square of 3/8 × 3/8 in (1 × 1 cm) squares. At each intersection of the lines, drill a hole passing through the board to the other side. Make sure the drill bit you use is only slightly smaller than the thickness of the nails. If not, the wood can split when you insert the nails. Make sure you drill right on top of your intersections. Now insert the nails. Before any tests, you need to make a frame, for safety. Nail the 1 5/8 in (4 cm) wide wooden strips onto the board to create frame edges (see figure).

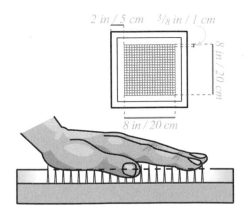

2 in / 5 cm   3/8 in / 1 cm

8 in / 20 cm

8 in / 20 cm

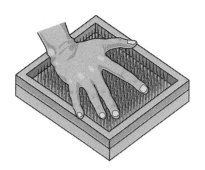

TESTS

1. Carefully feel the bed of nails with your palm.
2. Press a balloon filled with air against the point of just one nail.
3. Now carefully press the balloon against the entire bed of nails.
4. Place the bed of nails on top of a chair. Go ahead and carefully lower yourself down on it!

FUN FACTS

Each leg of a chair or bed carries only a quarter of the chair's weight. If you keep adding more legs, the weight is distributed over all of them, with each one "feeling" less weight each time (7 legs → 1/7 of the weight, etc). Imagine this chair full of legs, but upside down. That's your bed of nails!

An isolated nail could do you harm, like any other pointed object or sharp edge, when pressed against you, because the area where the force is applied is simply too small. That is not the case with a bed of nails or with a chair, for instance, since your weight (force) is then distributed over a much wider area. Each point within this area will exert a very small force on you. **Pressure** is a measure of the distribution of force over the surface where it is applied. If you double the area of the surface, you have to double the force to get the same pressure. You can express this by the relation Pressure = Force/Area. Can you now see why a fakir can sleep comfortably on his bed of nails?

How about putting the strong side of rulers to the test?

SUPPLIES

- 3 identical combs with a distance between the teeth large enough to fit a ruler
- 25 to 30 (thin, flexible) plastic rulers (enough to fit in the spaces between the teeth of a comb)

STEP BY STEP

Place two combs the same distance apart as the length of the rulers, then the third one centered between the two. Fit the rulers between the teeth of the combs one by one, as the picture shows. Remember Experiment 2, "How the Weak Become Strong"? Put the rulers to the ultimate test—your own weight. Sit on top of this set of rulers and discover how many it takes to hold you up.

FUN FACTS

As in the case of the bed of nails, your weight is distributed over all the rulers, which here are in their best condition to stand your weight. Since the bed of rulers has plenty of rulers to support you, you can even stand on it. (Just in case, it is wise to start with all rulers fitted in, and then remove them one by one.)

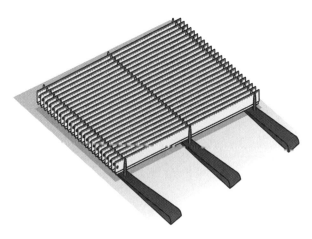

A submarine can both submerge and rise to the top of the water. What's the trick?

SUPPLIES

- 2-quart (2-liter) plastic bottle
- balloon
- plastic tubing, approx. 3 feet (1 m) in length
- silverware (3 or 4 table knives)
- kitchen sink or pail

Version 1

STEP BY STEP

Make two holes roughly 1 to 1 1/2 in (3 to 4 cm) in diameter in the upper and lower parts of the bottle, as shown in the figure. Insert one end of the tubing in the opening of the balloon. (It needs to fit snugly; use masking tape if necessary.) Place the knives (ballast) inside the bottle through the holes in the side. Insert the end of the tubing with the balloon into the mouth of the bottle and position the balloon in the middle of the bottle. Your new

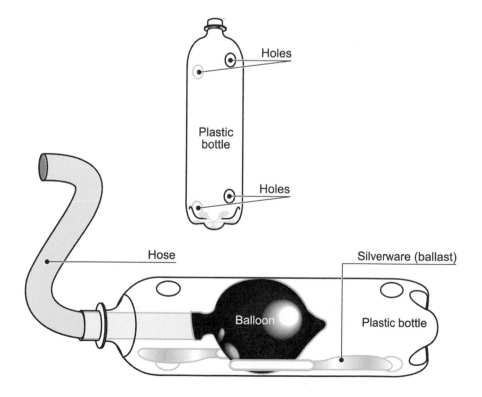

Holes

Plastic bottle

Holes

Hose

Silverware (ballast)

Balloon

Plastic bottle

submarine is ready. Put it on the bottom of the sink or in a pail. Fill the sink or pail with water until it covers your submarine. In order to not waste water, test your submarine when you are washing dishes in the kitchen sink. Blow into the free end of the tubing and see what happens.

FUN FACTS

If there were water in the balloon instead of air, the mass of the water would be about a thousand times greater than that of the air contained in the balloon. **Density** is a measure of the quantity of matter (mass) per unit volume. We can then say that water is a thousand times denser than the air in the balloon. When the balloon fills, the submarine rises so that more water—higher density—will remain at a lower depth in the sink or pail, and the air—lower density—will remain above. In this way, more matter (mass) will remain closer to the Earth's surface. This is a result of the gravitational attraction exerted by the Earth. Otherwise, we would simply float! This happens when we are in a swimming pool, for example, and our lungs are full of air. As soon as we let the air out of our lungs, we start sinking, just as happens with the "submarine" when we let the air out of the balloon.

Version 2

STEP BY STEP

Remove as much air as you can from the balloons. (If necessary, roll them as if they were almost empty tubes of tooth paste.) Insert the coins in the openings of the balloons.

SUPPLIES

- 2 party balloons
- 2 coins
- 2 plastic bottles, 1 large, 1 small, both with screw-on caps
- masking tape
- water

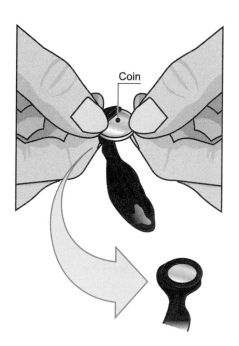

Coin

(The coins act as valves.) Put the balloons in the bottles, fill the bottles with water, and put the caps on tightly. The small amount of air that remains in the balloons will make them float, as shown. Ask someone to press one of the bottles while you press the other. Discover which of the bottles permits you to make the submarine submerge more easily.

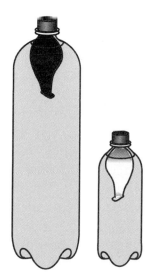

A. Squeeze

Tightly cap an empty plastic bottle, then squeeze it as hard as you can with your hands. You will certainly be able to deform it (at least temporarily). Try to deform a bottle filled with water, as in the experiment with submarines. The small amount of air in the bottle can be compressed, to a certain point. The water cannot, and for this reason it is said to be incompressible.

B. Blow

Make small holes in a plastic soda bottle in various places with a pin or needle. Blow through the mouth of the bottle and feel the air coming out of the holes with your fingers. Fill the bottle with water and cap it well. (If the holes are small enough, hardly any water will come out of them.) Squeeze the bottle. (Be careful! Water will squirt out of all the holes in all directions, like in a shower.) With this experiment, you show that pressure propagates in all directions in a fluid, including air (you can repeat the experiment with the bottle empty).

FUN FACTS

The submarine experiments demonstrate that any externally applied pressure is transmitted undiminished to all parts of an enclosed fluid (for example, air or water). Considering that the two plastic bottles in version 2 have different surface areas, when you squeeze them with the same force, you produce a different pressure in the water inside the bottles. This extra pressure, in its turn, forces the balloons to shrink, thus increasing effectively their average density. Eventually, when the applied pressure is large enough, the "submarines" will sink. Now, which bottle will sink more easily?

SUPPLIES

- plastic tube from a pen, with one end capped/sealed
- small vial, such as from a sample of perfume, or a dropper
- wooden matchstick
- plastic bottle with screw-on cap
- water

STEP BY STEP

Fill the tube (submarine) with water until it is almost filled, leaving only a small air bubble at the top. Fill the bottle with water and place the submarine inside so that it is suspended vertically. Close the bottle tightly and then press its sides with both hands, as the picture shows. What about using now the wooden matchstick as a submarine?

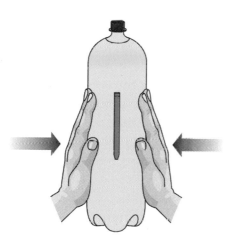

FUN FACTS

This version is just a variant of the previous model. In the present case, when you squeeze the bottle, the extra pressure produced inside the plastic tube squeezes the air bubble, shrinking it, hence increasing the average density of the submarine. In a real submarine, the effect you produced with your hands is made with suction pumps. They pump water into ballast tanks, effectively increasing the average density of the submarine until it submerges. To surface, they pump the water back out into the ocean, decreasing the average density of the submarine until the submarine rises to the surface. Although ships are made of materials denser than water, they don't sink. This happens because most of the ship's space is filled with air. If the ship's weight plus its load exceeds a certain value, the ship will inevitably sink, just as the submarine does with its ballast tanks full of water. Now, what happens if the bottle is not completely filled with water or if it has a small hole at its top or at its bottom? Will the submarine work the same way?

23

WATER AMPLIFIER (WATER TRANSISTOR)

It's easy to turn a little water into a lot of wine.

SUPPLIES

- two 2-quart (2-liter) plastic bottles
- plastic tubing, 1.6 ft (50 cm) in length
- cardboard or wooden box
- superglue
- water

STEP BY STEP

Cut one of the 2-quart (2-liter) bottles about 2 3/4 in (7 cm) below the neck to give you a "reservoir" (the bottom) and a funnel (the top). About 2 in (5 cm) from the bottom of the reservoir, make a round hole just big enough for the plastic tubing to pass through. Pass the tubing through the hole and bend it as the picture shows. Carefully use superglue to seal the hole around the tubing. Fill the reservoir so that almost the whole hose is underwater, as the illustration shows. Make a hole in the cardboard to fit the top of the bottle, upside down, like a funnel. Make another hole in the box for the plastic tubing to come out. Place the reservoir inside the cardboard box and close it. Place the cardboard on top of a table, near the edge. Place the hose that comes out of the box inside the opening of the second bottle, which is below the table. Pour a little water into the funnel, just enough to make the tubing completely submerged in the reservoir. Now try it and see how the trick works.

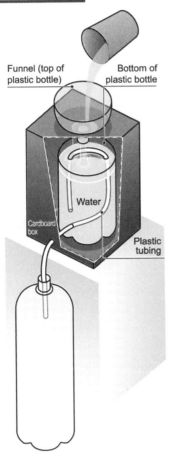

Funnel (top of plastic bottle)

Bottom of plastic bottle

Water

Cardboard box

Plastic tubing

HINTS

1. Try the experiment without the cardboard box and follow what happens. You can paint or color the box to make it look cooler.

2. You can amaze your friends by turning water into wine: just add some color, like grape juice concentrate, in the water of the reservoir.

FUN FACTS

When you lift an object a certain height, it acquires *gravitational potential energy*. This kind of energy is related to the Earth's gravitational attraction. When you release the object, it is accelerated downward so that the potential energy is converted into energy of motion, called *kinetic energy*. The tendency of everything, including water, is to be as close as possible to the Earth's surface in order to minimize its potential energy. (Water seeks its own level, hence the siphon effect.) In the water transistor, one end of the tubing is in contact with water at the bottom of the container, while the other end is almost at the same level but is in contact with air. The submerged part of the tubing is filled with water, as you can easily see. As you pour extra water into the reservoir, the water level in the tubing moves up, keeping at the same level as the top of the reservoir. When the tubing becomes totally submerged, water from the reservoir will start flowing down into the bottle, where its gravitational potential energy is minimized. Can you conceive other kinds of water transistors using different siphon systems?

24

HYDRAULIC ELEVATOR

Discover how to raise a heavy object with only a little force.

SUPPLIES

- Two pistons (see "How to Make Pistons out of PVC Pipes" printed at the end of this experiment), one with a smaller plunger (PVC pipe with external diameter around 1 in/2.5 cm) and the other with a larger plunger (PVC pipe with external diameter around 1 1/4 in/3.2 cm)
- pieces of wood with 3/8 in to 5/8 in (1 to 1.5 cm) thickness to make the support of the elevator (height = 8 in/20 cm, base = 4 in/10 cm, bench length = 8 in/20 cm)
- plastic tubing (aquarium-type), 8 in (20 cm) length
- nails or screws
- water

STEP BY STEP

Make two round holes at the top of the bench and in the fixed rack just below it, around the sizes of the smaller and larger pistons. Make a moving rack, with two finger-sized holes for the plastic tubing, to hold the two pistons tightly so that they cannot move vertically as their plungers move up or down. Connect the ends of the pistons to the plastic tubing. Pull out the plunger of the larger piston and fill it up with water.

Pressing the plunger down, force the water to fill up the whole length of tubing and the smaller piston. Press down on the plunger of the smaller piston. What happens? Now press down on the plunger of the larger piston. What is the relationship between the force necessary to move the water and the internal diameters of the pistons?

FUN FACTS

The hydraulic elevator works simply because water can transmit pressure efficiently, making possible a large multiplication of force. In the present case, that happens because the two pistons have different cross-sectional areas. (Compare the potential damage of an isolated nail with the harmlessness of a bed of nails, as in Experiment 20, "Bed of Nails.") The pressure you create by applying a force on a surface with smaller area (piston with a

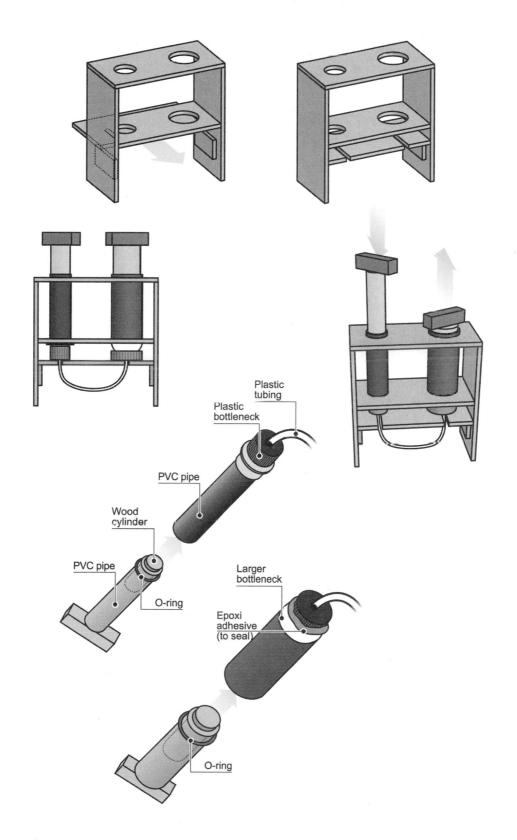

Plastic
tubing

Plastic
bottleneck

PVC pipe

Wood
cylinder

PVC pipe

O-ring

Larger
bottleneck

Epoxi
adhesive
(to seal)

O-ring

Fun with Mechanics

79

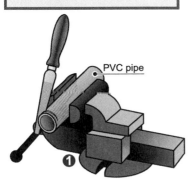

SUPPLIES

- 2 pieces of PVC pipe of equal length but different diameters so that one can slip inside the other, for example, the larger pipe with internal diameter equal to 7/8 in (2.2 cm) and the smaller one with an external diameter approximately 1/16 in (1.6 mm) smaller
- round (cylindrical) piece of wood with about 1 1/2 in (4 cm) in length to insert tightly into the smaller pipe
- O-ring for sealing when the smaller pipe is inside the larger one (you can find O-rings in hardware stores)
- candle
- piece of plastic tubing (aquarium-like)
- screw

PVC pipe

①

smaller cross section) is thus transmitted undiminished to the plunger of the piston with a larger cross section. This means that a larger force (pressure multiplied by area) will make the plunger with the larger cross-sectional area move up. Since water is an incompressible fluid, the volume of water expelled from the first piston by the depressed plunger must be equal to the volume of water received in the other piston by the movement of its plunger. What happens if the tubing or some other part of the system is expandable like a balloon? What if it leaks?

HOW TO MAKE PISTONS OUT OF PVC PIPES

Insert the round piece of wood entirely into the end of the smaller pipe. If necessary, use superglue to fix it and seal the end of the pipe. With this piece of pipe fixed, use a flat file tilted to make a notch around the pipe, about 1/4 in (6 mm) from the end of the pipe with the piece of wood inside (1). The notch should be regular and sized to accommodate the O-ring so that when the smaller pipe is inserted inside the larger pipe, the O-ring will provide a good seal. You may eventually need to insert another piece of wood in the opening of the smaller pipe with a handle fixed in it (fasten the piece of wood to the PVC pipe with a screw). Make a hole in the middle of the screw-on cap to snugly insert the end of the plastic tubing. (The hard plastic disk inside the plastic cap is essential for a good seal). Now, use a candle flame to slightly melt one end of the larger pipe (4). Then force it into the inner end of the bottleneck of a 2-qt (2-l) plastic bottle so that the end of the pipe stays about 1/4 in (6 mm) from the outer

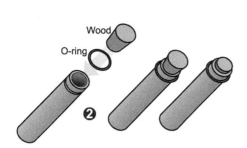

Wood

O-ring

②

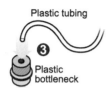

Plastic tubing

③

Plastic
bottleneck

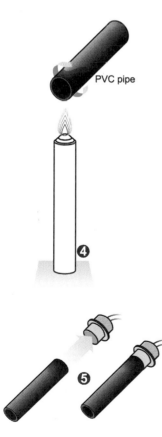

PVC pipe

④

⑤

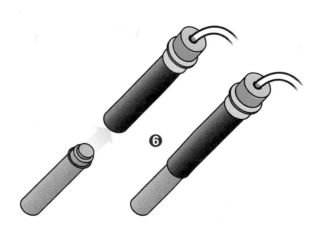

⑥

opening of the bottleneck. (You want to be able to screw the plastic bottle cap with the plastic tubing onto the bottleneck so that everything seals properly.) Remove the soot from the end of the pipe and then use superglue to fasten the end of the pipe to the bottleneck and to obtain a tight seal (5). Now, you just need to insert the smaller pipe with its covered end up (the "plunger") into the opening of the larger pipe to have your piston ready. In the case of a larger pipe with external diameter around 1 1/2 in (3.8 cm), you can use, for example, the bottleneck and cap of a liquid yogurt plastic bottle. In this case, you may have to use epoxy adhesive to seal the cap (see figure).

25

HYDRAULIC ROBOTS

Build a robot that can accomplish various tasks. You can even invent other models.

SUPPLIES

- 4 pistons, similar to the piston used in Experiment 24, "Hydraulic Elevator," but with a smaller diameter
- 3 wood rectangles roughly 1 1/4 × 3 1/2 in, 3/4 × 6 1/4 in, and 3/4 × 7 1/2 in (3.2 × 9 cm, 2 × 16 cm, 2 × 19 cm), 1/2 in (1.3 cm) thick
- square wooden base roughly 8 × 8 in (20 × 20 cm), 1/2 in (1.3 cm) thick
- hinge or piece of flexible plastic
- long screw with nut
- PVC pipe with external diameter slightly larger than the

pistons' diameter, enough for two 2 in (5 cm) sections
- 2-quart (2-liter) or 1-gallon plastic bottle with screw-on cap
- plastic tubing
- 2 3/4 in (7 cm) of solid wire about 1/16 in (2 mm) in thickness
- very small wood screws (to fasten the cap to the wooden base)
- wood screws or nails (for the control panel)
- small metal hook
- adhesive tape
- water

STEP BY STEP

Tube Holders for the Pistons

You will need two: one for the vertical (support) piston and one for the horizontal (on wood base). Cut two equal sections of the PVC pipe 2 in (5 cm) long. See inset 1 in the illustration on page 85 for the vertical piston holder. As the picture shows, take one PVC section and drill two holes all the way through from one side of the pipe to the other, about 3/4 in (2 cm) from each other.

Make the outside holes big enough for the heads of the screws to pass all the way through, but make the inside holes smaller so the screws will hold (the screws will attach the PVC section to the wood).

The horizontal holder (the second PVC section) needs only one pair of holes (all the way through) like the ones you made in the other section, with the outside hole big enough for the head of the long screw to pass through. Follow inset 2 of the illustration: put the long screw through the two holes, then attach the nut to the side under the PVC section. Fit one of the pistons into the horizontal holder with the plunger all the way in. Do the same for the vertical holder, using another piston. The pistons should fit snugly, under pressure, into both holders. (You can use adhesive tape around the pistons to get the snug fit).

Robot's Arm

Cut or saw the neck of the plastic bottle just below the lip of plastic on the neck of the bottle. (You might need to sand it flat.) Using two wood screws, attach the 3/4 × 6 1/4 in (2 × 16 cm) wooden rectangle (the vertical arm) to the bottle cap, as inset 3 shows (it may be necessary to drill two small starting holes). This gives you a rotating vertical support for your robot. Attach the small hook to the end of the 3/4 × 7 1/2 in (2 × 19 cm) wooden rectangle—the horizontal arm—and join the vertical support and horizontal arm pieces with the hinge or piece of flexible plastic, as inset 4 shows. Now attach the two-hole PVC holder to the vertical support so that the horizontal arm rests on top of the piston plunger perpendicular to the other arm (90°). Use two small wood screws, as inset 3 shows, to attach the neck of the plastic bottle to the center of the wooden base in the position indicated by the illustration (it may be necessary to drill small starting holes). When the neck is firmly attached, screw on the cap (with vertical support on top) about halfway, so that it is firmly attached but can easily swing in both directions from approximately the position shown.

Position the horizontal holder and piston with its top 1 1/4 in (3.1 cm) from the vertical support, forming a 90° angle. Now drill a hole in the wood base for the long screw to enter and fasten the single-hole PVC holder with the long screw and nut. This hole should be such that the holder can turn freely (see inset 2). Bend the wire 3/8 in (1 cm) from the end, making an L-shape. Drill a hole for the wire to enter in the wood piece inserted in the plunger of the horizontal piston, near the top. Also, drill a horizontal hole all the way through the vertical support for the wire to pass through, parallel to the base (to do this, unscrew the bottle cap from the bottle neck attached to the wood base). Now, pass the wire through the vertical support and bend the other end of the wire to fit it through the hole in the wood piece inserted in the plunger, as the main illustration shows.

Control Panel

Drill two different-sized holes in the 1 1/4 × 3 1/2 in (3.2 × 9 cm) wooden piece to snugly fit two pistons. Now attach this control panel to the wooden base with screws or nails.

How the Robot Works

Connect the tubing of the pistons in the control panel to the robot's arm. Fill the pistons with water, as you did in Experiment 24, "Hydraulic Elevator." You simply press down on the pistons of the control panel to make

Hydraulic Robot

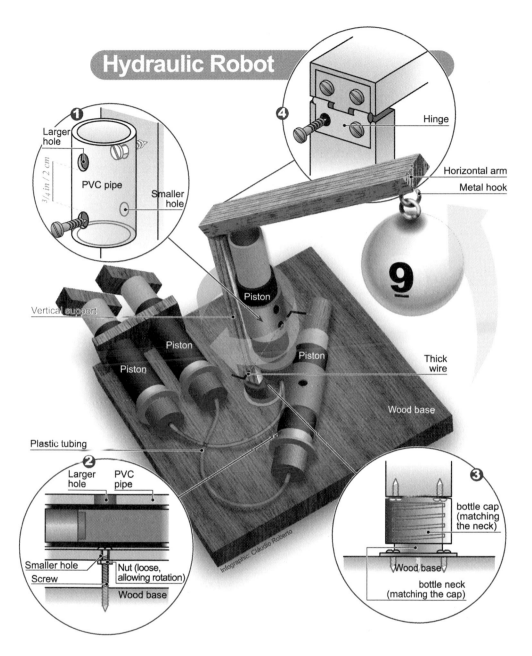

①
Larger hole
3/4 in / 2 cm
PVC pipe
Smaller hole

④
Hinge
Horizontal arm
Metal hook

9

Piston
Vertical support
Piston
Piston
Piston
Thick wire
Wood base
Plastic tubing

②
Larger hole
PVC pipe
Smaller hole
Screw
Nut (loose, allowing rotation)
Wood base

③
bottle cap (matching the neck)
Wood base
bottle neck (matching the cap)

Infographic: Cláudio Roberto

your robot work. Don't use your robot to raise objects exceeding 1 pound (500 grams).

> **HINT**
>
> How about including small levers ("force multipliers") to actuate the plungers of the pistons of the control panel?

The hydraulic robot is based on transmission of pressure and hence of force through a liquid. In our model, pressure is used to rotate the base and the robot's arm. All hydraulic devices, such as mechanisms for opening bus doors, controllers of flaps in the wings of airplanes, car brakes, compressors, jack hammers, elevators, jacks, and many others, including more complex equipment, are based on the very same principle of the simple hydraulic devices you can build using basic items. Isn't that cool?

26

DRAWBRIDGES

Make a pair of drawbridges using the same principle as the hydraulic elevator and robots.

SUPPLIES

- 3 pistons, like those used in Experiment 25, "Hydraulic Robots"
- narrow plastic tubing (aquarium type)
- aquarium T-joint for plastic tubing
- PVC pipe
- nails and wood screws
- 2 hinges or 2 pieces of flexible plastic
- wooden rectangles

STEP BY STEP

Attach a hinge or piece of flexible plastic to the top end of each of the two column pieces (the vertical wood rectangles in the illustration), using screws or nails. Make two piston holders (fixed, with two holes) according to the instructions in Experiment 25. Place the pistons in their holders and mark where to screw the holders onto the columns at the right height so that the hinges rest horizontally on top of the plungers (when all the way down), at a 90-degree angle to the vertical columns. Remove the pistons and screw the holders to the columns. Now attach the hinges or pieces of flexible plastic to the hor-

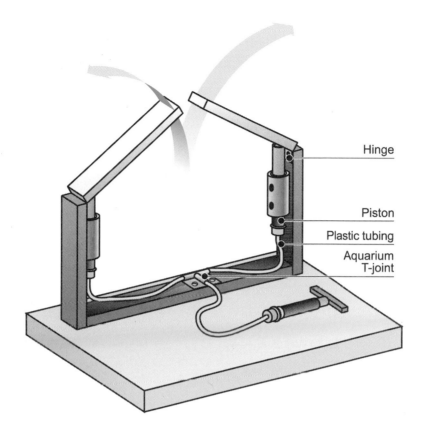

izontal bridge pieces, and reinsert the pistons in the holders. Nail or screw the column pieces to the wooden base, the length of which is slightly longer than the horizontal bridge pieces placed end to end. In the center of the base piece, attach the aquarium T-joint. Connect plastic tubing between the points (the "injection ends") of the pistons and the two sides of the T-joint. Connect the final piston (the controller or actuator) to the middle of the T-joint. Fill the hydraulic system (the tubing and control piston) with water so that the column piston plungers are all the way in (down) and the control piston plunger is all the way out. Pressing in the actuator (plunger) of your controller (piston) will raise up the two sides of the drawbridge

Labels in figure:
- Hinge
- Piston
- Plastic tubing
- Aquarium T-joint

simultaneously. Pulling the actuator plunger out again will lower the drawbridge, allowing foot traffic and vehicles to pass over it.

FUN FACTS

As you press the actuator plunger, the force you apply displaces the water and thus transfers the applied pressure to both pistons attached to the column pieces of the bridge. Their plungers, in turn, move up, forcing the moving sides of the bridge to swivel upward. Based on the principle that pressure propagates in water in all directions, you can conceive many other ingenious devices (for instance, more sophisticated hydraulic robots). Are you ready for that?

27

CIRCUMVENTING OBSTACLES: HOW AIR AND WATER STREAMS FIND THEIR WAY

Bending air streams produces a lot of cool effects, from juggling balloons to sprays to the lift of airplanes and more. Here we explore just a tiny bit of this fascinating universe with simple yet intriguing experiments.

SUPPLIES

- hair dryer (cold air)
- vacuum cleaner
- small balls of cotton or paper
- drinking straw
- glass of water or can of soda
- adhesive tape

A. From High Pressure to Low Pressure

STEP BY STEP

Place the nozzle of the hair dryer on a smooth flat surface (floor or top of a table) to produce a directional flow along the surface. Put a small ball of cotton or paper in front of the air output of the hair dryer and switch it on, using the cold air setting. Replace the hair dryer with the vacuum cleaner, putting the small ball about 2 in (5 cm) from the air intake nozzle. Last, but not least, drink the water or soda using the drinking straw to suck it up.

FUN FACTS

The small balls (your "probes") were initially at rest before you switched on the hair dryer or the vacuum cleaner. The water in the glass was also at rest. One ball is kicked away by the air flow coming out of the hair dryer while the other is sucked in by the vacuum cleaner. The water is sucked up by your mouth. The balls (and hence the air) are thus accelerated when a difference in air pressure occurs. They start with zero speed and end up with a finite speed. The same holds for the water (and the air initially present in the straw) you suck up with the drinking straw. In all these cases, you can say that a difference in pressure produces a force that accelerates the air and water. Can you relate the air pressure (where it is higher and lower) with the corresponding speeds of the air flow produced by a difference in air pressure (where the speed is higher and where it is lower)? In the next experiments, we explore how differences in air pressure can also bend air and water streams. To begin with, let's check how a flat table affects an air stream coming out of a hollow pen tube or drinking straw positioned at different angles. Feel with your hand what happens with the air stream after it comes out of the tube or straw. You can use a small tent made of paper to probe the air pressure, as shown in the figure. Make sure that the tent's lateral walls are kept fixed; you can use adhesive tape to fix the walls to the surface of the table. (Hint: the air pressure above the tent is atmospheric pressure.) If the tube or straw is tilted, how would you expect the air stream to move after it hits the surface? (Imagine a ball hitting a wall.) What happens if the tent's roof sinks? What do you notice when the air stream crosses the tent parallel to the table's surface? Does sucking air produce the same effects as blowing air?

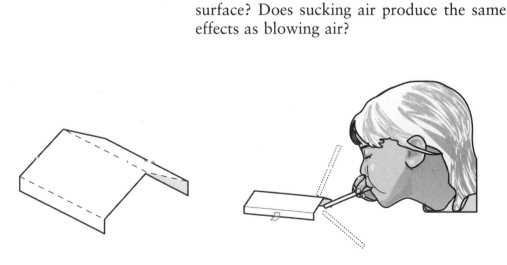

B. Party Balloon Faces Water and Air Flows

STEP BY STEP

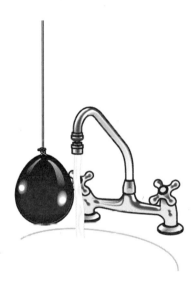

Attach one end of the string to the neck of an inflated balloon. Hold the other end of the string and let the balloon hang below and just to the side of the tap (see figure). Open the tap completely so that the water coming out hits one side of the balloon. How does the water flow along the surface of the balloon? What happens to the balloon? Now attach the free end of the string to a heavier object (away from the sink) so the balloon won't fly away, and use the hair dryer (cold air) to blow an air stream vertically down along the side of the balloon, as shown in the figure. Try also a horizontal air stream aimed at the top of the balloon. What happens to the balloon? Compare the effect of the air and water streams on the balloon.

FUN FACTS

As a directional air or water stream hits a surface, it drags along the air as the stream moves along the surface, creating a partial vacuum beneath the stream, which forces the stream to bend and follow the surface. This is called Coanda effect. On the other hand, the pressure on the side of the balloon away from the air or water stream remains at atmospheric pressure. The pressure imbalance thus created makes the balloon moves sideways. What happens when the horizontal air stream hits the top of the balloon?

C. Race Competition

This experiment should be done on a smooth surface, such as a floor, and in a closed room with air conditioning, heat, and fans switched off unless otherwise suggested.

STEP BY STEP

Place the output of the hair dryer close to the glass wall with the nozzle resting on the floor. Put two small pieces of cotton or two small balls of paper of different colors (your "probes") close to the glass rim, as shown in the figure. Turn on the hair dryer and see which of the cotton balls or crumpled paper moves faster. Use your hand and pieces of cotton or paper positioned around the glass to further explore what happens to the air flow after it hits the glass. You can draw on a piece of paper the trajectories of the small balls placed at different positions around

the glass to visualize the air flow. (This might be easier if you have a camcorder to film the experiment; you can play it back in slow motion.) What about placing a filled party balloon close to the glass wall at the position of the colored piece of cotton shown in the figure? How is the air pressure at this spot? Does a difference in air pressure occur? Can you now discover the relation between air pressure and the speed of the air flow around the glass? What makes the bending of the air flow possible? You can also replace the glass with a bucket, the hair dryer with a fan, and the small balls with ping-pong balls. What about replacing the hair dryer with the vacuum cleaner (the air is now sucked instead of blown!) Enjoy yourself with this thrilling race!

FUN FACTS

At the spot where the air stream coming out of the hair dryer hits the glass, the air pressure becomes higher than the atmospheric pressure. As the stream moves around the glass surface, its velocity changes direction and the air stream is accelerated. These experiments demonstrate that changes in pressure produce the forces necessary for the air flow to bend and to speed up. (Eventually, an air flow can encounter higher air pressure. What happens then?) Can you find out where the air pressure is lower compared to atmospheric pressure and relate this difference to the speed of the air flow at various positions around the glass?

28

JUGGLING BALLOONS

Do you think you can keep three or more balloons moving in circles in the air, using only a hair dryer? That's the challenge!

SUPPLIES

- 3 party balloons
- hair dryer (cold air) with base to hold it up

STEP BY STEP

Place the dryer upright in the holder so that the cold air coming out goes straight up. Blow up the balloons and tie a knot in the end so the air can't escape. Place the balloons in the air stream one at a time, so that they are held up by the air flow. It may be necessary to tilt the hair dryer a bit (see the picture).

FUN FACTS

The air stream coming from the hair dryer hits the balloons and, in their turn, the balloons hit the air back (action = reaction). When the air stream coming up vertically hits a balloon right above it, the air stream moves up along the surface of the balloon. This means that the air flow that was originally directional must bend sideways. In this case, all sides of the balloon force the air stream to bend, so no net effect should occur. When one side of the balloon moves away from the center of the air stream rising from the hair dryer (extreme right and left in the figure), an imbalance is produced. The air pressure at the side less exposed to the air stream is the atmospheric pressure, while a partial vacuum is created at the opposite side due to the air stream. In this way, the balloons tend to stay bouncing around near center stage! Now, why do they bounce around there rather than

remain still at the center of the stream of air? Can you figure out what will make them bounce around faster? Slower? Farther? What about switching the hair dryer setting between low and high?

29 AIR STREAMS ON TOP OF CARS, ROOFS, AND MOUNTAINS

Discover what happens on top when a lateral stream of air turns up.

SUPPLIES

- empty pen tube or drinking straw
- piece of paper 3 × 6 in (7.5 × 15 cm)
- glass (preferably clear)
- small piece of cotton or crumpled paper
- bucket
- inflated party balloon (smaller than the bucket)
- fan

STEP BY STEP

Bend the middle of the paper around a finger to make a bump there (your "mountain" or the top of a car). Hold the paper on a surface as shown in the figure. Place one end of the pen tube or straw close to the bump and blow through the other end. What happens to the top of the bump? Since the pressure in the "tunnel" below the bump is atmospheric pressure (there is no air flow there), what do you conclude about the pressure on the top of the bump? What about the speed of the air flow there (place a small ball of cotton or crumpled paper to "probe" it.) Does the air flow speed up as it moves up to the top? Now, place the glass with the small piece of cotton or paper inside on top of a work surface (see Experiment 30, "Make Your Own Sprayer," next). Position the output of the hair dryer in front of the side of the glass with the nozzle resting on the work surface. Move the nozzle up with the hair dryer switched on and see what happens when the air flow starts crossing the top of the glass (see figure). Now

replace the glass with the bucket with the balloon inside, and replace the hair dryer with the fan. Position the fan so that the air flow it produces is below the top of the bucket. Move the fan up and see what happens. You can cover the bucket with a page of a newspaper (the "roof") to see the effect of a lateral wind on it.

FUN FACTS

To circumvent an obstacle, the incoming air stream must change direction. This requires a force component perpendicular to the direction of the flow. As the stream hits the obstacle, it is slowed down and the pressure increases locally (kinetic energy is transformed into "pressure energy"). Consider the air pressure around the obstacle at points away from the spot directly hit by the incoming flow. If the pressure at these points becomes less than the atmospheric pressure, a net force is created perpendicular to the air flow, which enables it to bend. In addition, the pressure imbalance around the obstacle speeds up the air flow as it moves toward the point of lowest pressure. Pressure energy is then transformed into kinetic energy (Bernoulli's effect). Imagine a flag that is spread out in the wind with a ripple in it. The air pressure over the ripple decreases so that the air stream can bend as it crosses over the ripple. The air pressure on the two sides of the flag near the ripple becomes unequal, just as in the experiment with the piece of paper. As the ripple moves randomly all over the flag, the flag flutters in the wind.

MAKE YOUR OWN SPRAYER

Sprayers are used with perfume bottles, spray paint, and bug spray, for example. Make your own sprayer and find out how it works.

SUPPLIES

- cardboard
- drinking straws
- adhesive tape
- glass of water

STEP BY STEP

Cut the cardboard in a fat L-shape, as the picture shows. Tape the pieces of straws to the cardboard so that the opening of the horizontal straw is partially blocked by the vertical one. See the inset blow-up of the figure. Place the end of the vertical straw in a glass of water and blow into the horizontal straw. Your sprayer is ready. You can replace the horizontal straw with one slightly larger and partially block its air opening with the vertical straw as you did before. Now what happens?

FUN FACTS

The sprayer is just a variant of the previous experiment. The vertical straw represents the glass (or bucket), and the water corresponds to your probe: the piece of cotton inside a glass (or a balloon inside a bucket). The air flow coming out of the end of the horizontal straw is partially blocked by the vertical straw. It plays the role of the air flow hitting the glass (bucket) when you switch on your hair dryer (or fan). As the water arrives at the top, it is split into tiny droplets and taken away by the air

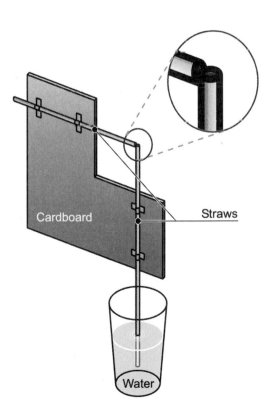

Cardboard

Straws

Water

flow. What would happen if, instead of blowing, you suck the air? Would the sprayer work the same way?

A STEP FURTHER

Can a vacuum cleaner go mad?

STEP BY STEP

Attach one end of the string to the balloon's neck. Hold the other end of the string and place the balloon close to the air input of the vacuum cleaner. Do you notice something strange in the balloon's behavior as it hits the vacuum cleaner nozzle? Aren't vacuum cleaners meant to suck things in?

FUN FACTS

The filled balloon has a flexible surface. When you stretch it, it stores elastic energy. If you release the balloon, it will move around. (Its elastic energy is converted into motion. Just try it!) Now, if you put the palm of your hand less than 3/4 in (2 cm) from the air input of the vacuum cleaner, you will definitely feel an air flow. (Would you feel it if your hand were further away?) As the space between your palm and the air input decreases, the air speed increases (you can also use light objects to "probe" it) until it stops altogether when you block the air input completely. What about the air pressure at the nozzle when air is sucked in? Does the air flow produce an extra vacuum? What happens to the pressure inside the nozzle when

- vacuum cleaner
- inflated party balloon
- piece of string 1 1/2 ft (45 cm) long

the air input is completely blocked? Since the sucking power of the vacuum cleaner does not change, can the vacuum cleaner keep the balloon stretched at a pressure lower than it can "afford"? With so many clues, you will surely find out what is behind this weird yet cool effect.

31 WIND TUNNEL

Isn't it amazing that an airplane, much heavier than air, can fly? Find out the importance of the wing in keeping it aloft.

SUPPLIES

- piece of paper
- fine thread pieces or small (but not tiny) paper strips
- white glue or adhesive tape

A. Airplane Wing

STEP BY STEP

Hold a sheet of paper (the "wing") with the fingertips of both hands and blow over the top of the paper, then blow on the bottom side, as the picture shows. Check out what happens in each case. To see how air flow acts on this wing, glue or tape small bits of the thread or strips of paper at different spots on the top and underside of the paper. Watch the bits of thread or paper as you blow. What happens to the air flow when it touches the wing? How does the wing react?

FUN FACTS

When you blow the piece of paper from above, as indicated, the air flow must bend

to follow the paper. This, in turn, requires that the air pressure on top of the paper decreases. Since the pressure on the other surface of the paper is the atmospheric pressure, the paper is lifted. When you blow the piece of paper from below, the air flow must be accelerated downward to move along the paper surface. This means that the air pressure below the paper increases in relation to the atmospheric pressure. Hence the paper is pushed upward (see the next Fun Facts for another interpretation.)

B. Wind Tunnel and Angle of Attack ★

STEP BY STEP

Cut the 2-quart (2-liter) plastic bottle, giving you a tube around 9 in (23 cm) in length and open on both ends (the wind tunnel). Bend the solid wire into a two-pointed, square-U-shaped fork, as the picture shows. Cut two slots 2 3/4 in (7 cm) long and about 1/8 in (3 mm) wide—parallel to each other and to the central axis of the tube—from the top part of the tube. Also cut two holes in the

---

SUPPLIES

- 2-quart (2-liter) plastic bottle
- construction paper
- solid wire
- glue
- drinking straw
- electric fan (or a good set of lungs)

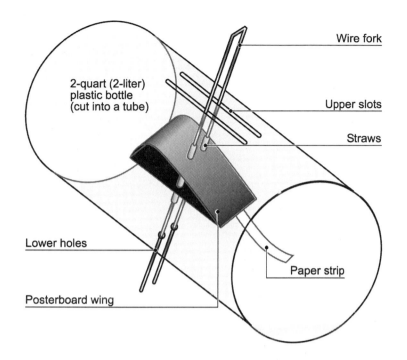

Wire fork

2-quart (2-liter) plastic bottle (cut into a tube)

Upper slots

Straws

Lower holes

Paper strip

Posterboard wing

underside of the tube to fit the wire fork into, following the illustration. Make your wing from the construction paper, 2 3/4 in (7 cm) wide and 2 3/4 to 3 1/4 in (7–8 cm) long, in the shape of an airplane wing (see the illustration's red wing). Make four holes in the wing for the wire fork, two above and two below, slightly smaller than the pieces of drinking straw, then fit the drinking straw pieces all the way through the wing. Now, carefully push the wire fork through the slots on top of the wind tunnel, through the straw pieces in the wing, and out through the holes at the bottom of the wind tunnel. This makes it possible to change the tilt of the wing (angle of attack) by just moving the fork back and forth. Place the tunnel opening in front of a fan and see what happens to the wing when you change the angle of attack. You can also place little pieces of thread or small strips of paper (as in the previous experiment) and gradually increase the angle of attack by moving the wire fork very slowly. This experiment shows how the wing can keep the airplane aloft by the action of airflow even when the airplane is upside down. Add "flaps" fixed to your wing and see how they affect the wing's performance. Try other shapes of wings and find out what advantages their shapes may offer.

FUN FACTS

In the absence of turbulence, a horizontal air stream hitting an object follows the contours of its surfaces (the Coanda effect). A prerequisite is that the speed of the air (and other fluids, like water) is not too high. Other factors should also be taken into account, like the density or viscosity of the fluid, which is a measure of the difficulty of a fluid sliding past itself—compare honey and water—and geometrical factors; otherwise, turbulence will occur, disturbing the smooth flow. Since the lower surface of the wing is flat, the air stream bends only if the wing is tilted. In that case, the air goes down the lower surface as it moves along the wing profile. This implies that the air is accelerated downward as it hits and moves along the wing. The air stream then slows down, and the air pressure increases at the lower surface of the wing. As a reaction, the wing experiences an upward force, known as lift. The upper surface of the wing produces a lift even if the lower surface is horizontal. In this case, the air flow moves up and then down along the wing profile. This requires a lower pressure at the top of the wing. The air flow is thus accelerated as it moves up (see Experiment 29, "Air Flow on Top of Cars, Roofs, and Mountains,") and when it is deflected downward as it leaves the wing. As a result, there is a net lift perpendicular to the wing. The horizontal component of the net lift is a drag force, known as *induced drag*. The air stream extracts energy from the airplane as it is deflected, slowing down the airplane and creating the drag effect. (Replace the fork with two pieces of string to hang the wing and check if the wing is dragged along with the streamline.) As the angle of attack increases (the wing is more tilted), turbulence becomes important. In this case, the air stream over the top of the wing separates from its surface. A turbulent air pocket forms above the wing, making the airplane unstable. (The lift then decreases and the drag increases.) You can see the effect of turbulence with the small bits of thread or strips of paper attached to the wing in the wind tunnel.

C. Pressure Map ★★

STEP BY STEP

SUPPLIES

- 1-pint (1/2 liter) plastic bottle
- construction paper
- 2 cardboard rectangles roughly 4 × 10 in (10 × 25 cm)
- Styrofoam board
- 1.8 ft (55 cm) of solid wire
- 2 to 3 push pins
- 2.3 ft (70 cm) of transparent tubing with inner diameter around 3/16 in (4 mm)
- 1.5 ft (45 cm) of flexible tubing with inner diameter around 1/8 in (3 mm)
- water
- food coloring (for contrast)
- hair dryer with base to hold it tilted
- adhesive tape
- superglue
- art/craft knife

Make a cup out of the bottle about half its height and a holder from the remaining part of the bottle. Cut a piece of transparent tubing 4 in (10 cm) long. Fold the solid wire, making an external frame out of it for the remaining piece of the transparent tubing, as indicated in the illustration on page 106. Fasten the tubing to the external frame using adhesive tape. Make a small disk of construction paper 1/2 in (1.26 cm) in diameter and carefully superglue it on one end of the 4 in (10 cm) piece of transparent tubing. (This will form the main part of the pressure probe.) Connect the other end to the flexible tubing. Carefully punch a small hole in the center of the disk. (This will be the sensitive end of the probe.) Make a wing from the construction paper, 2 3/4 in (7 cm) wide and 2 3/4 to 3 1/4 in (7–8 cm) long (see the illustration for the previous experiment, "Wind Tunnel and Angle of Attack"). Copy the wing profile twice on the Styrofoam board, as indicated. With an art/craft knife, make two identical pieces of Styrofoam that fit into the wing and insert them laterally. Fold the cardboard rectangles 2 in (5 cm) from one of the 4-in edges, forming two L-shaped panels 8 in (20 cm) high to attach to your work surface and hold the wing. Cut a rectangle from the Styrofoam board 2 3/4 in (7 cm) wide and insert it between the two panels at the base of the Ls (to act as a separator and to keep them perpendicular to the work surface). Fasten their base (the bottom parts of the Ls) to your work surface with adhesive tape. Insert the wing between the two panels and attach it to them with the push pins and a piece of adhesive tape. (The tape is to keep the tops of the panels together; see illustration.) Then place the hair dryer in front of the wing, as indicated. Submerge the curved end of the framed transparent tubing into the cup containing water with food coloring. Fasten the other end of the framed tubing on top of the holder with adhesive tape. (Adjust the height of the holder by turning the cap so that the straight part of the framed tubing is kept tilted slightly upwards). The probe will allow you to map out the pressure distribution on both the upper and lower sides of the wing when the hair dryer is switched on (using cold air). In order to do so, you first need to suck up the water until it fills the tubing to the point indicated by zero so that the level of the water in the cup and in the framed tubing is kept equal. (Remember to check this and keep the pressure probe zeroed in between measurements.) Then connect the flexible tubing to the main part of the probe. Keeping the small disk parallel to wing surface, switch on the hair dryer (cold air) and observe the variations of pressure along the upper and lower surfaces of the wing. To better visualize the changes in the water level in the transparent tubing, attach a strip of paper to the tubing at the place indicated. On which parts of the wing is the air pressure lower, and where is it higher? Considering your pressure map, can the tilted wing keep the airplane aloft?

CAUTION ————————————————

Be very careful when using an electric appliance, such as the hair dryer, near water. Do not put the hair dryer in the water. Do not spill water on the hair dryer. Do not put your hand in water or stand in water that might have spilled on the floor while you are holding the hair dryer.

FUN FACTS

There is a special reason why the probe suggested here to detect pressure variations along the wing surfaces requires a small disk with a small hole in its center. Without the disk, the air flow that is being probed would be bent by the probe, much in the same way as happens in a sprayer (see also Experiment 29, "Air

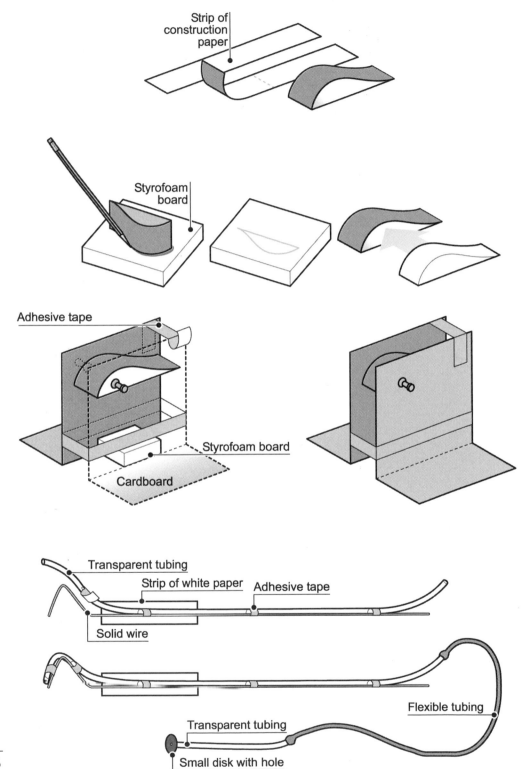

Strip of
construction
paper

Styrofoam
board

Adhesive tape

Styrofoam board

Cardboard

Transparent tubing

Strip of white paper

Adhesive tape

Solid wire

Flexible tubing

Transparent tubing

Small disk with hole

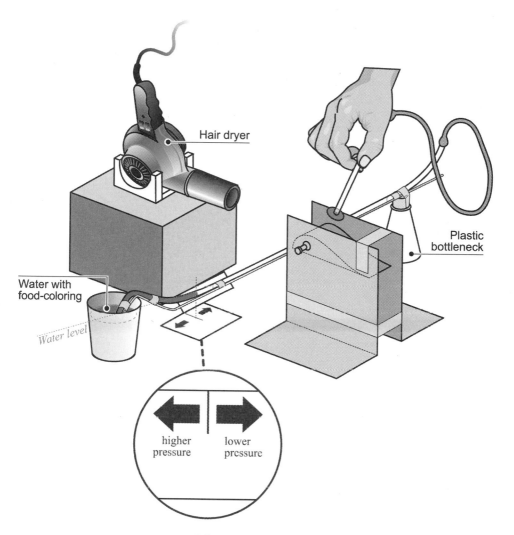

Hair dryer

Plastic
bottleneck

Water with
food-coloring

Water level

higher
pressure

lower
pressure

Flow on Top of Cars, Roofs, and Mountains"). The small disk prevents this from happening, thus leaving the probed air flow practically undisturbed. Notice that in the figure, the disk is parallel to the wing. Now place the disk close to the nozzle of the hair dryer and parallel to the air flow. Is the pressure indicated different from the atmospheric pressure ("static pressure")? Now place the disk perpendicular to the air flow, first closer to, then away from, the air output. What happens now, and why? Can you calibrate your pressure probing system (manometer) using standard pressure units?

UNWANTED BALL

A simple ball of crumpled paper can be kept out of a bottle even when you try hard to blow it inside.

SUPPLIES

- plastic bottle
- ball of crumpled paper

STEP BY STEP

With the bottle lying on its side, place the little ball in the opening, resting inside the neck of the bottle. Try to blow the ball inside the bottle by blowing on the opening. Now try again, but blow from the side. Finally, punch several holes in the bottle and repeat the experiment.

ALTERNATIVES

Instead of a bottle, use a tube of paper with only one opening. Try different sized balls. Try bottles or tubes with different sized openings and different ratios of the size of the ball to the size of the opening. If you punch holes in the side of the bottle or tube, does the size of the holes make a difference?

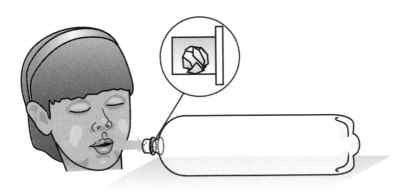

HINT

Check out Experiment 24, "Whirlpools (3D Vortices)" under *The World of Atoms and Our World: Cold, Heat, and Giant Bubbles.* Can you see how these two experiments are related?

FUN FACTS

Although air can be compressed (see Experiment 22, "The Submarine," under "A Step Further," part A), it is still made of matter, which needs space. (Tires, full balloons, and balls are all good examples of air's need for space.) As you try to blow the small ball into the bottle, it is kicked back, for there is no space available for the extra air you are blowing in. When you punch a hole in the bottle, some air can go out, which makes space available for the ball.

33 OUTSMARTING FRICTION (FLYING SAUCER)

Can you beat friction, the "sticking force" that prevents us from sliding when we walk? Discover the principle of hovercrafts that makes these vehicles float over the floor and over water.

SUPPLIES

- CD or small plastic disk
- balloon
- plastic bottle cap
- superglue

STEP BY STEP

Make a hole in the center of the cap. Superglue the cap's top to the center of the CD, as shown. Blow up the balloon and twist its opening so that the air can't escape, but don't knot it. Stuff the balloon's end into the cap. Place the disk over a smooth surface and release the balloon.

FUN FACTS

Friction happens when two surfaces in close contact "grip" each other. If the surfaces are rough, they anchor mechanically in each other (the "peaks" of one surface fit into the "valleys" of the other surface). If the surfaces are flat and smooth, friction arises due to short-range forces between atoms at both surfaces that act over nanometers (one nanometer defines the length scale of the nanoworld and is about 70,000 smaller than the diameter of a human hair). The harder the surfaces press together, the stronger they grip. When the surfaces are no longer in contact, friction disappears. The short-range forces between the ground and your feet, for example, are a consequence of electrostatics (see the experiments under *Electrifying Experiments: Electricity and Magnetism*) and quantum mechanics, the physics of the nanoworld—so standing is a kind of "nanolevitation."

The air pressure provided by the balloon produces a slightly pressurized air cushion underneath the CD. (You can also demonstrate this action by placing a CD on a flat, smooth surface and blowing air at the hole in center of the CD. Use a drinking straw or an empty pen tube held vertically with one end close to the surface. The air flow will find its way beneath the CD.) The pressurized air cushion causes the CD to rise or lift, and friction between the CD and the working table decreases enough that the CD can easily slide across the table. Since the air escapes sideways underneath the CD, the CD remains lifted or "on a cushion" only if there is a continuous flow of air. The lift will stop when the balloon becomes empty. It is therefore essential to keep the cushioning air from escaping. This is achieved by the use of a "skirt," which contains the air. Also, the total amount of weight that a real hovercraft can lift is equal to cushion pressure multiplied by the area of the hovercraft. Let's tackle these challenges in the next two models.

A STEP FURTHER ★

STEP BY STEP

Trace a hole around the opening of the hair dryer, and make the hole in the wooden disk with a diameter a bit smaller than that of the air outlet of the hair dryer, as shown in the illustration. Find some PVC pipe with an inside diameter that will fit the air outlet of

SUPPLIES

- wooden disk, 1/4 to 3/8 in (7–10 mm) thick with a radius of 5 1/8 to 5 1/2 in (13–14 cm)
- plastic bag (make it a strong one)
- small, hard plastic disk, like the inside of the cap of a plastic bottle
- PVC pipe
- small nails
- insulating tape
- hair dryer
- superglue

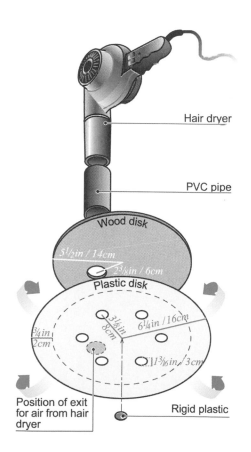

Hair dryer

PVC pipe

Wood disk

5½in / 14cm

2⅜in / 6cm

Plastic disk

¾in / 2cm

3⅛in / 8cm

6¼in / 16cm

1³⁄₁₆in / 3cm

Position of exit for air from hair dryer

Rigid plastic

the hair dryer. Cut a section of PVC pipe and fix it carefully with superglue on the wooden disk centered on the hole. To produce a skirt, cut a circle from the plastic bag with a 6 to 6 1/2 in (15–16 cm) radius. Stretch its edges out and cover the bottom of the disk, wrapping the edges around the edges of the wooden disk and sealing them very well on the top of the disk with the insulating tape so that air can't escape. Now take the hard plastic disk and with a nail right in the center, attach it to the bottom of the wooden disk. (This will keep the plastic from bulging and contacting the floor in the middle.) With the art/craft knife, cut out six circles from the plastic disk with about a 1/2 in (1.5 cm) radius, where the air will come out. Check out the illustration. Fit the end of the hair dryer into the pipe so that it will stay upright on the wooden disk, with its handle close to the center of the disk. (If you put an extra weight on top of the disk, you may need to turn the handle 180 degrees and move the weight around until the disk becomes balanced.) Then turn the hair dryer on (cold air) with the disk lying on top of a table.

CAUTION ————————————————

Do not try to fly your saucer on water because of the danger of electrical accidents.

Try placing various weights on top of the disk to see how your flying saucer really manages to outsmart friction.

SUPPLIES

- wooden disk roughly 5/8 in (1.6 cm) thick and with a radius of about 1.4 ft (42.7 cm)
- plastic covering for the bottom of the disk (like vinyl)
- bottom of a plastic bowl about 10 in (25 cm) in diameter, or a metallic disk the same size and 1/16 in (2 mm) thickness
- electrical tape
- screw with nut and washer
- vacuum cleaner that also operates as a blower with extension hose
- hose as long as the perimeter of the wooden disk
- wood screws (to attach the hose to the disk to protect its border)

How big is your sense of adventure? Want to make a flying saucer that can lift over 500 lb (250 kg) of weight? Follow the same steps as in the previous experiment. Scale measurements according to the ratio of the diameters (or radii) of the disks. Use adhesive tape (insulating tape) to seal the plastic covering around the disk top and a washer between the screw and the bottom of the plastic bowl or a metallic disk placed at the center of the bottom of the disk. Cut the air holes in the plastic, scaling them as suggested above. Drill a hole in the wooden disk that fits the attachment of the vacuum hose, also scaling the distance between the centers of the hole and the wooden disk as suggested. If it's not a tight fit, seal it with insulating tape. To protect the edge of the disk, drill through the hose and use screws to attach it to the disk edge (circumference) all the way around. To do this, make evenly spaced holes all the way through the tubing, as in detail inset 1 in Experiment 25, "Hydraulic Robots."

FUN FACTS

Under the saucer, a pressurized volume of air is formed in the skirt so that the saucer floats above the ground. The skirt is kept pressurized by the air coming from the hair dryer, which compensates for the air leaving through the six holes in the plastic bag attached to the saucer. Since friction starts acting only when the bottom of the saucer is in contact with the ground, the saucer can easily move around. As soon as the hair dryer is turned off, the skirt empties, the saucer stops floating, and friction comes in. Vehicles based on this same principle can move on the ground or on water.

> **CAUTION**
>
> Do not try to fly your saucer on water because of the danger of electrical accidents.

A wider saucer will require more power to keep the air bag full of air, but since the air pressure in the bag balances the pressure exerted by the weight of the saucer and its cargo, which now is distributed over a wider area, it can support much more weight.

WHEEL THAT ROLLS UPHILL

Common sense says that wheels roll downward. You can build one that rolls upward.

SUPPLIES

- large metal can (like a juice can)
- magnet, like those found in speakers, or a battery (or other object with the same weight and dimensions)
- masking tape
- flat piece of wood

STEP BY STEP

Stick the magnet on the bottom of the can, near the side, or tape the battery solidly to the side of the can. The ramp can be made by simply leaning the piece of wood on a book.

HOW IT WORKS

Place the can at the foot of the ramp, so that the magnet or battery is at the top of the can. When you let the can go, it will climb the ramp, but meanwhile, what happens to the magnet or battery?

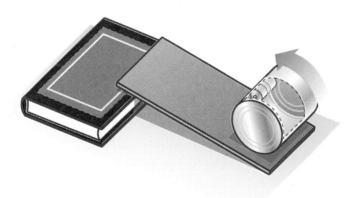

SUPPLIES

- two 2-quart (2-liter) plastic bottles with conic necks and screw-on caps
- 2 thin pieces of wood (like restaurant chopsticks) or thick, straight wires (for example, from a coat hanger)
- masking tape

Cut the two bottles about 3 1/4 in (8 cm) below the opening, making two identical funnels. Make sure the cut edges are nice and even. Use the masking tape to make the "wheel" shown in the illustration, whose "axle" is formed from the two bottlenecks. Draw a dot at the center of one of the caps (axis) with a marking pen. Place the two chopsticks in a V-shape, with the top ends held higher using, for example, a book. Place the wheel inside the V at the bottom. Before releasing the wheel with its axle parallel to the ground, determine the height of the dot (axis) using a ruler or a caliper. When the wheel stops rolling, determine the new height of the dot. How is the dot's height relative to the ground compared to its initial position? Does the wheel really climb?

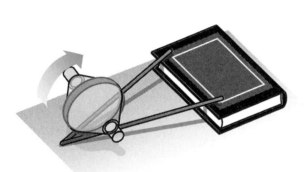

FUN FACTS

A solid body can be viewed in terms of a point where all its mass is concentrated, called the *center of mass*. Since the tendency of everything is to be as close as possible to the Earth's surface in order to minimize its potential energy, the center of mass of the wheel will inevitably fall down as the wheel climbs up. In a ring, for example, the center of mass is located at its center, although no mass actually exists there. To see this, imagine the circle divided into a large, even number (like 1000, 1002, 1004, and so on) of equal small pieces. You can always find two small pieces diametrically opposed. The center of mass of these two small pieces can only be at the center of the circle (where else could it be?). The same holds for the whole ring. This same reasoning can be applied to the can. The weight attached to it changes the position of the center of mass, which moves closer to the ground as the can rolls uphill. In the case of the double-funnel wheel, you can imagine it as a set of small rings. Because the two funnels are equal, the center of mass of the wheel must be at the center of the circle where the two funnels meet. Now, what about placing the chopsticks parallel to each other? Will the wheel climb up this time? Try also rolling the wheel with the axle down the incline, starting from the top. How fast does the wheel's axis move down? Would a wheel with a different geometry (a cylindrical pencil, for example) have better performance? What about a race competition involving two equal cylindrical plastic bottles, one empty and the other completely filled with water? Let both roll down a ramp at the same time. Which one moves faster? Why?

HINT

As the wheels roll down, their gravitational potential energy is converted into motion. You may notice two kinds of motion: rotation and translation, which is what really matters for this race competition. If a wheel's rotational sluggishness (inertia) is greater, where will most of its potential energy go?

Have you ever noticed that when ballerinas spin around, they first open their arms and then hold them close to their sides? Why do they do that?

STEP BY STEP

Sit down, stretching out your arms and legs. Ask someone to spin the chair. Pull your arms and legs in and discover the trick of the ballerina.

A STEP FURTHER

Hang a ruler vertically by a string attached to one end and try to spin it (around its vertical axis). Then tie the string around the middle of the ruler and let it hang horizontally. Which way does the ruler spin more easily?

FUN FACTS

Who has a harder time walking: an elephant or an ant? The mass of a body determines the difficulty it will have in starting to move in a straight line or to stop (translational inertia). In contrast, the inertia associated with rotational movement (called "moment of inertia," the *rotational mass*) depends on the mass of the body *and* its geometric distribution, taking the rotation axis as a reference (see this experiment's "A Step Further"). From this perspective, with arms and legs stretched out, you are an elephant "dancer"; but with them pulled in, you become a dancing ant. To achieve this effect, you have to make some effort. When you pull in your arms and legs, your kinetic energy increases, hence your higher spinning speed. As long as you are not subjected to any major external influence, your tendency is to keep spinning forever. As you decrease your rotational inertia, you have to preserve something to compensate it. Physicists call this *angular momentum*, which is the product of the rotational mass of an object and its spinning speed. So, as your rotational inertia decreases, your spinning speed must increase. If you love ice skating, you probably have already experienced this cool effect. Can you recognize it in other situations?

Have you ever noticed that you don't fall off a bicycle when you're pedaling? Try sitting atop one with it stopped! And when you go around a curve, why do you lean? All of this is easily explained.

SUPPLIES

- bicycle wheel with its spokes, hub, and axel (look for a used one at a bike repair shop)
- 2 wooden handles
- 6-sided nuts for the axle
- epoxy (hardening kind) or superglue
- rope or strong twine, 1 to 2 ft (1–2 meters)

STEP BY STEP

Drill a hole in each handle to fit the nuts in, as the illustration shows, and glue them in securely. When the glue has dried, screw the handles onto the ends of the axle. Tie one end of the rope to one of the handles, and tie the other end high above, such as to the roof or a tree limb, or have someone hold the end securely, but so that you can reach it easily. Hold the free handle securely, spin the wheel, and let go of the handle. What happens? Why don't you fall off a bicycle when you're pedaling? Grab the free handle and try to change the direction of the axle.

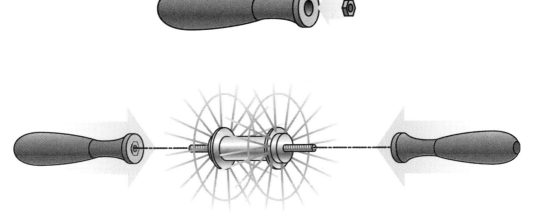

A STEP FURTHER

A. Spinning Coin

Stand a coin on its edge. Try to knock it over by blowing on it. Easy, isn't it? Now, stand the coin on its edge and spin it rapidly, as shown. Now try to knock it over just by blowing on it. Is that easy? What similarity can you see between the coin and the wheel when they are spinning? Try rolling a ball forward over a hard surface. Does it balance itself as it moves on?

B. Leaning Wheel

With the wheel spinning, grab both handles and lean the wheel, such as when you ride your bicycle around a curve. When the leaning is combined with the curving, is it easier? What happens with the spinning axle in this case?

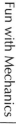

Fun with Mechanics

If nothing disturbs a spinning body, it will continue to spin forever. To change this condition requires external action, as the previous experiments demonstrate. It is a consequence of the conservation of angular momentum, a quantity of motion associated with spinning objects (see also Experiment 34, "The Ballerina's Trick"). Spinning generates stability due to rotational inertia, called the gyroscopic effect. You can also see the gyroscopic effect in action with your air propulsion rocket (Experiment 15, "Rockets with Chemical and Air Propulsion"). Because of the gyroscopic effect, cement mixers can topple over on sharp curves if the mixer is spinning too fast. (The rotational inertia of the spinning mixer tends to keep its axis of rotation at a fixed direction. The spinning speed of the mixer must then be kept very low to avoid accidents.)

You can determine the direction of the angular momentum of the spinning wheel (held by the rope attached to one handle) using a very simple rule. Point the curved fingers of your right hand around in the direction of the wheel's rotation. Your thumb will then point to the right, which is the direction of the wheel's angular momentum. (If you use the left hand, your thumb will point in the opposite direction; this means that the angular momentum does not have a mirror image.) The center of mass of the spinning wheel, which coincides with the hub's center, is not straight under the rope that supports the wheel. If the wheel were motionless, it would simply fall down in order to minimize its gravitational potential energy. The center of mass would then tend to stay just below the rope's end. This does not happen because the wheel is spinning. However, gravity's twist tends to lean the wheel to the right side (see figure). Using the right hand rule, you can determine that at the position of the wheel shown in the figure, gravity produces an extra component of angular momentum pointed toward the rear. The angular momentum then drifts toward the right rear. The twist exerted by gravity keeps adding new components to the angular momentum, perpendicular to the wheel's axis and parallel to the ground. As a result, the wheel's axis rotates counterclockwise around a vertical axis coincident with the rope. This motion is called gyroscopic precession.

If you are riding a bicycle and lean your body to the right, the bicycle will steer toward the right, as happens with the suspended wheel. As you lean and turn to make a curve, an additional force is necessary to change the direction of your motion, so you end up with two forces: gravity and friction between the tires and the ground. The resulting force is aligned with the center of mass of the system consisting of your body and the bicycle. This force exerts no twist in the same way as pressing a door against its hinges does not make the door rotate. The gyroscopic precession and the turning of the bicycle about its steering axis thus work together to make a forward-moving bicycle incredibly stable.

When you are in a car or bus and it accelerates or brakes, what do you feel? Build your own acceleration gauge to demonstrate situations like these and the effects of acceleration in curves.

SUPPLIES

- Styrofoam ball or cork (as from a wine bottle)
- clear glass or plastic jar with screw-on top
- string
- 1 pint (1/2 liter) plastic bottle with cap
- superglue
- epoxy putty (the kind that you can shape like clay, and that hardens)

CAUTION

Read directions for use of epoxy. Use rubber gloves; latex gloves are not recommended. Work in an area with good ventilation. Use only soap and water to wash hands after use.

STEP BY STEP

With superglue, carefully glue one end of the string onto the Styrofoam ball or cork. Use the epoxy to stick the other end to the center of the inside of the jar lid. Fill the jar with water and screw the lid on tightly, with the ball or cork inside. Place the jar upside-down and the ball will float in the water. Your acceleration gauge is ready.

ALTERNATIVE

Make a carpenter's level (water-level) using a watertight plastic bottle, full of water, with one air bubble. Does the bubble work like the Styrofoam ball or cork?

TESTS

You can take your acceleration gauge into a car or bus and see what happens when you round a curve, or speed up or slow down

going in a straight line. You can also place your acceleration gauge on top of a chair or lazy Susan and test it. To avoid accidents, make sure you attach it solidly to whatever surface you have chosen. You can also use your acceleration gauge to find out more about your own walking rhythm.

FUN FACTS

When you are sitting on an accelerated platform (car, bus, plane, or a spinning chair, for example), the acceleration of the platform produces a local gravitation (see Experiment 11, "Astronaut in the Elevator"), which is in the opposite direction of its acceleration. As heavier (denser) objects tend to go to the "bottom" (in the case of a car, bottom can mean either right or left or to the side, depending on whether it breaks, accelerates, or makes a curve), the parts of the accelerometer less dense than water (Styrofoam ball, cork, or air bubble) move in the direction of the acceleration. The accelerometer thus allows you to demonstrate that the more accelerated a car is, the greater is the gravitation produced in its interior, as in an accelerated spacecraft. You can also hold the accelerometer while sitting on a spinning chair. Keep it close to your body and then stretch your arms, so that you still can see your "probe" (ball, cork, or bubble). You can thus use the accelerometer as an alternative to the "Stretching Carrousel" (see Experiment 6) to determine the acceleration of a spinning object attached to a string when the string's length is increased and the same spinning speed is maintained.

38

RAW OR HARD-BOILED EGGS

It's easy to find out if an egg is hard-boiled or not without having to break the shell.

SUPPLIES
- hard-boiled egg
- raw egg

STEP BY STEP

Using your fingers, stand a hard-boiled egg on its end and then stand a raw one on its end. Spin them like tops. What's the difference between the movements of the two eggs?

A STEP FURTHER

Standing with your arms a bit away from your sides and keeping your eyes closed, turn around slowly. After a few turns, try to stop dead or walk in a straight line. What do you feel?

In the raw egg, the shell is solid, but its interior contains a liquid (egg white plus yolk). When you spin the shell, the liquid in the interior of the egg is a bit sluggish in following your "order" (the liquid has rotational inertia). Thus part of the egg (the shell) tends to spin, while the other part tends to remain at rest. This conflict produces a "drunken" egg. To cure the egg, just boil it until it becomes hard. You can now spin it and the egg will follow your "order," since now it is consistently solid. Inside of our ears there is also a liquid that turns when we spin, telling the brain about our movement. When we stop, the liquid keeps spinning, just like our raw egg or water in a teacup. Try stirring a spoonful of sugar in a teacup in circles and then pull the spoon out—what happens to the water? To the brain, you are still spinning, although you have already stopped. Combine this with your body being stopped and you get weird sensations (dizziness plus what else?!). What if just after you stop spinning, you spin the other way around? Does this help in curing yourself of the weird after-spinning sensations?

39

HAND-OPERATED WATER PUMP (ARCHIMEDES' SCREW)

How can water climb while "descending"? Doesn't it seem contradictory? Make your own water pump and find the answer.

SUPPLIES

- PVC pipe 1.6 ft (0.5 m) long
- transparent plastic tubing, 1 yd (1 m) long
- masking tape
- 2 bowls
- water
- water-based poster paint or food coloring

STEP BY STEP

Twist the tubing around the PVC pipe and fix it in place with the tape, as the illustration shows. Fill one of the bowls with water (the pool) and place one of the ends of the pipe in it. The pipe needs to remain at an incline. The other end of the pipe should be higher up, leaning on the other bowl (the reservoir), so that the water from the tubing will empty into the reservoir. Hold the higher end of the PVC pipe and turn it with your hand. It will be easier to see what's happening if you put some coloring in the water in the "pool," perhaps water-based poster paint or food coloring, since they are safe and easy to clean up afterwards.

HINT

To follow up on what happens to the water when you operate the pump, just make a smaller version of it and insert a small ball or any other object into the tubing where it can move freely. By turning the PVC pipe, you can see what happens to the ball. Be prepared to collect it at the upper end of the tubing!

FUN FACTS

The tubing around the PVC pipe can be viewed as a collection of ramps wrapped around a central axle. As you turn the axle

Hand-operated water-pump
(Archimedes' screw)

PVC pipe
(1/2 yd/0.5 m)

Reservoir

Masking
tape

Water

Clear plastic
tubing (1 yd/1 m)

"Well"

Water

Infographic: Claudio Roberto

☀ HINT

Hold the upper end of
the PVC pipe and turn it with
your hand. It will be easier to see
what happens if you add a little
food-coloring to the water in the "well." Watch
the path of the water in the tube as you turn the
PVC and find out why it "climbs."

(PVC pipe), you carry water in successive lifts. As the water moves up, it is always on a ramp and "falls down" in order to locally minimize its gravitational potential energy. As you keep turning the wheel, the water moves on to a new ramp until it comes out at the top reservoir. The effort you have to make to pump water up with the Archimedes' screw is much smaller than that needed to lift the water vertically. It is just like walking on a road or path around a mountain up to the top. It requires less effort compared to climbing directly up the mountain, though the distance you have to walk is much longer. Now, try to lift water turning the axle (PVC pipe) the other way around. Will it work now? Why?

ALTERNATIVE: VERTICAL PUMPING ★

STEP BY STEP

Insert the 1.6 ft (0.5 m) tubing into the 2 in (5 cm) tubing (if necessary, use electrical tape to obtain a tight fit). Place the ball in the wider tubing, on top of the 1.6 ft (0.5 m) tubing. The ball will act as a valve. Next, insert the 1 1/2 in (4 cm) long piece of tubing into the wider tubing, leaving some space for the ball to move up and down (if necessary, use electrical tape to obtain a tight fit).

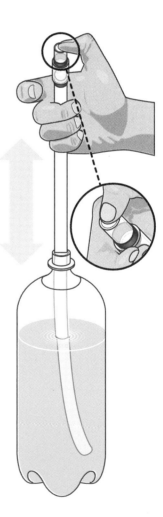

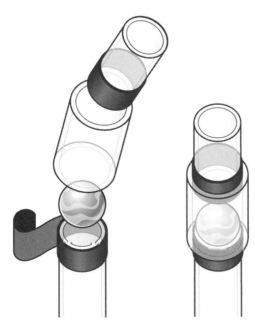

OPERATION OF THE PUMP

Submerge the longer tubing in the bottle filled with water and move it quickly up and down, without the bottom part coming out of the water. At the same time, keep the opening at the top of the tubing partially covered with your thumb. See if you can get the water to gush out of the upper opening. You can adjust the space available for the ball by pressing the 1 /2 in (4 cm) long piece of tubing or pulling it out. Watch the column of water form in the clear tubing. You can use food coloring to better see what is going on.

FUN FACTS

As you move the pump up, the ball keeps the top end of the longer tubing closed, while the air inside the tubing has more space—you produce a partial vacuum. Once the air pressure inside the tubing decreases, the water moves up (the same happens when you suck a beverage using a straw). When you move the tubing down, the ball is too sluggish to follow the tubing, so it lets the air out while more water comes into the tubing. By repeating this operation, more water fills the tubing until its only alternative is to gush out. Now, why do you need to partially cover the top of the pump with your thumb?

Make a simple rotating water fountain and find out why it rotates.

SUPPLIES

- plastic drinking cup
- 2 bendable drinking straws
- superglue
- string
- water

STEP BY STEP

Do this experiment over the sink or a basin to keep from getting the floor wet. Make one hole in the side of the cup near the bottom (see illustration). To do this, you can light a match, blow it out, and immediately press it to the side of the cup. Stick one straw inside the cup, as shown (1) and seal it with super-glue. Make two small holes at the lip of the cup, one on each side, to pass the string handle through. Tie a longer string to the middle of the handle. Holding the other end of the

string, fill the cup with water (while it is hanging suspended) and watch what happens. Did the fountain twist? Now, empty the cup and make the second hole exactly opposite from the first, then stick the second straw into it, as shown (2). Fill the cup again. What's the importance of having the two bent straws?

FUN FACTS

According to the action = reaction principle, when water comes out of a bent straw, the straw recoils just like a rotating lawn sprinkler. However, one stream of water might not be enough to overcome the cup's rotational inertia and the string resistance to being twisted. When you add an extra straw, as shown in the illustration, the two streams of water work together and produce the movement you observe. If air or water were sucked in, however, there would be no noticeable recoil in the opposite direction. Can you devise simple ways to demonstrate this surprising effect?

Have you ever noticed what is left over at the bottom of a jar of nuts as they are eaten over time? See why that happens.

SUPPLIES

- potato, onion, cork, egg, ping-pong ball, stone (try various objects)
- jar with rice, beans, marbles, sand, or nuts with different sizes (big and small mixed together)

STEP BY STEP

Place the chosen object (like an egg, for example) at the bottom of the jar and cover it with the rice, beans, or sand. Shake the jar up and down, as the picture shows. What happens? Try shaking the jar laterally. You can also have a go with the nuts before eating them up.

FUN FACTS

When you shake the jar up and down, both the big pieces and the small pieces are accelerated. As you move the jar down, the small pieces are quicker to find their way back to the space made available when the big pieces move up. (What happens if the density of the big pieces is greater than the density of the small pieces?) The fact that we have big rocks at the surface of the Earth might also be a consequence of the shakings up and down produced by earthquakes over the past billion years.

Granular matter displays a fascinating behavior. Try also another experiment. Push a wide stick into a pail of sand until it stops. Now, keep steadily pressing the stick into the sand and shake the pail. Does the stick start going into the sand more easily? Do you get the same result if you shake the pail laterally? To go deeper into the sand you need to make space for the stick. Which way is more effective? Why?

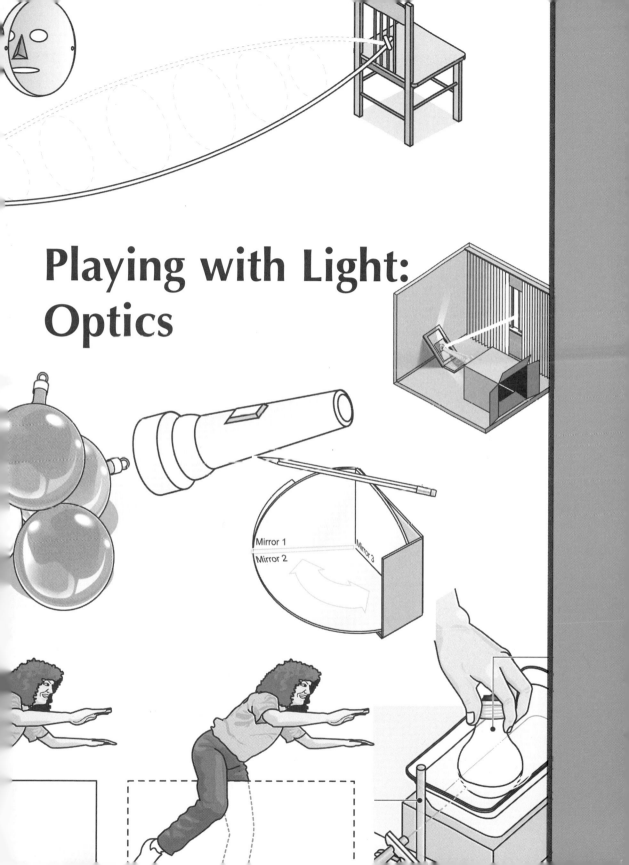

Playing with Light: Optics

INVISIBLE GLASS

See how you can make a bottle invisible!

SUPPLIES

- smooth, transparent drinking glass (not colored)
- 2 or 3 tubes of glycerin (100 ml, found at pharmacies)
- empty bottle of Tabasco, or other narrow transparent glass bottle, very clean and dry, with a screw-on top
- PVC pipe
- piece of paneling
- superglue

STEP BY STEP

Fill up the bottle with glycerin and screw the lid on tightly. Fill about one-third of the glass with glycerin. Slowly place the bottle in the glass. A part of the bottle will be submerged. What happens to this part? Why? What will happen if you place the bottle with glycerin in a clear glass with a third of its volume filled with water?

OPTIONAL

Use the PVC pipe, superglued onto the piece of paneling, as a holder for the glass. This will keep the glycerin from spilling over.

FUN FACTS

Everyone knows that placing a knife in a glass of water makes it look bent. Of course, the knife out of the water isn't bent. Check it out for yourself. The light that comes from each part of the knife reaches you at the shortest time possible. (Light is most efficient!) In water, the speed of light is less than in air or in outer space. (You can also express it in terms of the *refraction index*, which is the ratio of the speed of light in the outer space and the speed of light in a transparent medium, like water and glass.) So, light is better off if it takes a shorter path in water,

where it is slower (has a higher refraction index), and a longer path in air (has a refraction index of approximately 1), where it is faster. If light did not bend when it comes out of the water, its path in water would be longer and in air it would be shorter, so on the whole, the light would take more time to reach you. It then "prefers" to bend. This shows that light "sees" water and air as different media. What can you say now about the glycerin and glass? Do they look different for light? (Are their refraction indexes different or the same?)

| 2 | ★ |

DECOMPOSING LIGHT INTO A RAINBOW: 21ST-CENTURY VERSION OF NEWTON'S CLASSICAL EXPERIMENTS

Use two CDs to produce special rainbows.

SUPPLIES

- 2 CDs
- cardboard
- cardboard box
- black matte construction paper
- adhesive tape
- electrical tape (black)
- ordinary (converging) lens
- highlight pen

STEP BY STEP

Fold a piece of black construction paper to form a holder for one of the CDs, as shown in the figure. (Use adhesive tape to keep the fold fixed.) Attach the holder to a tilted board with adhesive tape. (If necessary, use a heavy book to keep the board from slipping.) Make a hole about 1 × 1 in (2.5 × 2.5 cm) at the center of the cardboard bottom. Make a shutter with strips of cardboard whose windows are two pieces of cardboard glued on top of slightly longer pieces of construction paper (see detail) with straight edges. Attach the shutter to the inner wall of the box with the hole so that the construction paper edges stay parallel to each other. Line the inside of

 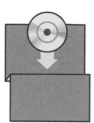 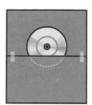

 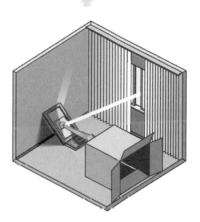

the box with the black construction paper. Position the board with the CD holder and the box so that a specific color hits the shutter. (Increase or decrease the shutter's opening as you choose to get more or less light coming through.) With the second CD, try to decompose the color singled out using a white piece of paper as your screen. Can you further decompose light? Was Newton right?

A STEP FURTHER

Attach a white piece of paper (your white screen) to the box, where the rainbow is formed. Try to encompass all colors within the lens (to decrease the size of the rainbow, move the box toward the CD holder). Change the distance between the lens and the white paper until you get a light point on it—this corresponds to the *focal distance*. (You may also have to change the inclination of the lens in relation to the screen.) Can you sum all the colors with your lens? Compare your results with the light point you get by focusing the Sun's rays directly on a piece of white paper.

> **CAUTION**
>
> Focusing the Sun's rays on a piece of paper can cause a fire, so do it carefully and for only a short time.

Use your lens to see if all the colors of the rainbow projected on the screen have the same focal distance.

With the highlight pen, draw a design on your hand. Select a color to come through the shutter and place your hand inside the box in front of the shutter. Which color best highlights the design on your hand?

FUN FACTS

A CD is just a piece of plastic about four one-hundredths (4/100) of an inch (1.2 mm) thick with a single spiral data track circling from the inside of the disc to the outside. The separation of one track from the next is only 1.6 microns (the diameter of a human hair is around 50–100 microns), and the whole track has tiny microscopic bumps 0.5 microns wide, a minimum of 0.83 microns long, and 125 nanometers high. (A nanometer is a billionth of a meter.) The tiny separation of one track from the next in a CD

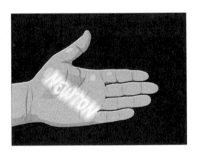

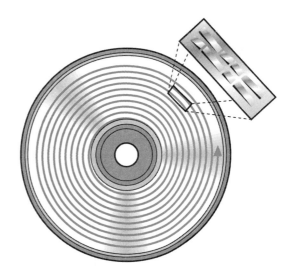

makes all the difference. As light passes around an edge or through a slit, it is bent. This bending is called *diffraction* and is related to the wave nature of light (see Experiment 13, "New Discoveries with Polaroids," in this part of the book). When the dimensions of an obstacle become comparable to the wavelength of the incident light, as happens in a CD, the usual reflection produced by mirrors is replaced by a scattering of the light by the obstacle, causing diffraction (see also Experiment 13, "Car Control by TV Control," under *Electrifying Experiments: Electricity and Magnetism*). You can check this action by replacing the CD with an ordinary mirror. As the light from the Sun hits the CD, its very fine spiral scatters the light components in different directions, hence the rainbow formed on the wall of the box. (A CD can thus be considered a super-prism.) New CDs with even more compact spirals are coming up on the market that can store much more data. Would they change anything in the proposed experiments?

LENS AND THE REFRACTION INDEX (LIGHT DISPERSION)

An ordinary converging lens is designed to focus the incident light on a screen by keeping the lens at a certain distance from a screen (for example, a piece of white paper). If the refraction index of the lens were the same for all components of the Sun's light, they would all focus on the screen at the same point. This would mean that the total time necessary for each incident ray to cross the lens and then travel in the air before it hits the screen would be the same for all colors. This would also imply that the lens refraction index would be the same for all components. Is that true or not? Check it out for yourself! Consider reflection in an ordinary flat mirror. Do all the components reflect the same way, or is there any "dispersion"?

> **HINT**
>
> Newton conceived a new telescope by replacing a lens with a curved mirror. Is there any advantage in that? You can think of a curved mirror as a collection of tiny flat mirrors.)

THE LIGHT THAT NEWTON NEVER SAW

The ink of highlight pens produces a nice effect called fluorescence. When you illuminate the design in a dark room with ultraviolet (UV) light, which is not visible, the ink emits visible light in a remarkable way. (The UV component comes after violet, the highest frequency of visible light.) So, be ready to experience something unique.

CHALLENGE YOUR PERCEPTION

Our visual perception of the world around us relies on the response of our eyes to light and on how our brain interprets the patterns we see. Let's put your eyes to the test.

SUPPLIES

- plastic mask of a face (available at party stores)
- light source to light it

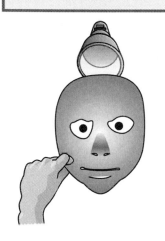

A. Living Masks

STEP BY STEP

Draw a black circle for each eye (pupils) on two small pieces of paper. Stick the eyes on the front side of the mask, as the picture shows. Light up the mask with a light source (like a flashlight or sunlight) so that the back side of the mask is facing you (see the picture). Step back from the mask and watch its eyes. Pay attention to what happens when you walk slowly sideways in front of the mask.

SUPPLIES

- mask of a face
- 2 elastics
- 2 paper clips

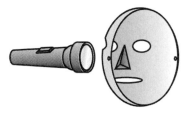

B. Frankenstein

What about rotating a mask? Make two small holes in the mask, one at its top and the other at its bottom. Pass one end of the elastics through each hole and attach it to a clip. Hold the free ends of the elastics with your hands. Ask someone to rotate the mask so that the two elastics become very wound up. Then release the mask and see what happens. Why not draw funny features in the hollow part of the mask with a marking pen to make it look different from the face? Can you create new Frankensteins?

FUN FACTS

Having two eyes allows us to see things in three dimensions. Our depth perception, however, is conditioned by our familiarity with the features of a face sticking out, not hollowed in. Also, when you look at the backside of the mask, the image arises from the fact that there are areas in the mask that are more illuminated and others that are less illuminated. When you move sideways, the pattern of light and shadow you see changes. This gives the impression that the mask is following you.

C. Bending Parallel Lines

Parallel lines never cross. Test the truth of this famous principle.

STEP BY STEP

Draw several parallel lines, evenly spaced apart. Based on the lines you have drawn, make squares until your drawing looks like the black-and-white figure shown here. Then color in the alternating squares, as the next figure shows. When you are done drawing, see what happens when you step back and look at it.

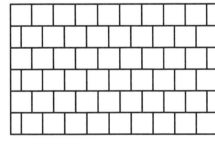

FUN FACTS

Rows of light and dark tiles appear wedge-shaped, depending on the brightness of the "mortar" between the tiles. The size of the tiles also matters—if they are too big, you may have to step back and stay at a certain distance from the tiles for the trick to work. Check it out for yourself! This optical illusion gives us some hints on how we see edges.

D. Black on White (Benham's Disk), Shrinking Objects, Newton's Disk, and the Crazy Roulette Wheel

You can make chemistry with colors.

STEP BY STEP

Use the set-up you built for Experiment 13, "Flattening the Earth at the Poles," in the *Fun With Mechanics* part of this book, but without the Earth and its axis (wire). Where the Earth was, place a screw. Tape a copy of the disk pattern of your choice on the smaller wheel. Turn the handle of the larger wheel, keeping your eyes on the pattern. In the case of Benham's disk (A) and the disk with a spiral (B), turn the wheel slower. In case (B), a few seconds later, look at the palm of your hand. Does it seem to shrink?

ALTERNATIVE

Cut a piece of cardboard the same size as the various disk patterns. Stick each pattern in turn on the cardboard, using tape or paperclips. Make a hole in the center of the cardboard disk and pass the string or twine

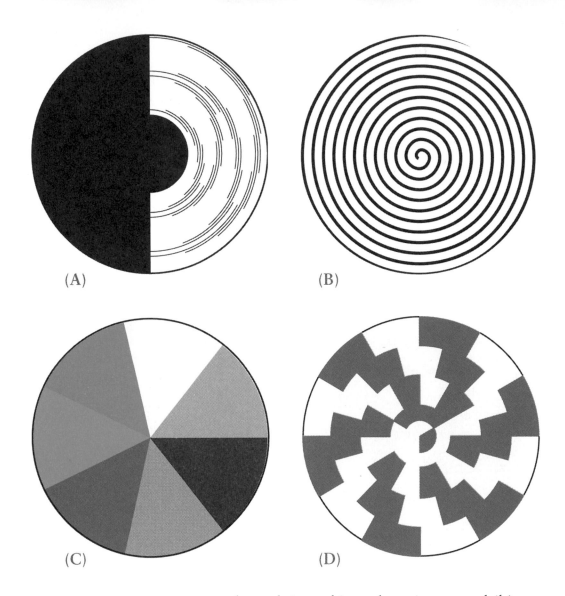

(A) (B)

(C) (D)

through it, making a knot in one end (bigger than the hole). Hold the other end of the cord and twist it until it is wound up tightly. Then let go and watch the pattern as it spins. From that point, follow the instructions of the previous experiment. You could also glue a ping-pong ball in the center of the cardboard disk, or stick a pencil through and spin it like a top (see the illustration).

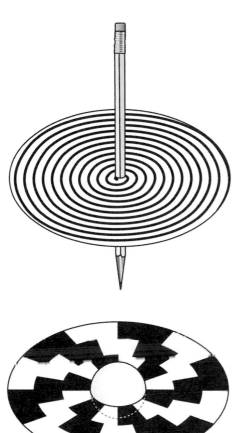

FUN FACTS

Each eye has small light receptors, called rods and cones. Each receptor is very specialized. Rods only detect differences between light and dark and are much more sensitive to light than cones, which detect color. When you look directly at an object, the light that enters your eyes falls on an area predominately made up of the less sensitive cones. Some types of cones respond more quickly than others, and some of them keep responding longer than others. If you look slightly to one side of something faint, it can be seen much better because the light from the object falls on areas of the retina (at the back of eye) that have many more rods. Each spinning pattern produces a special optical illusion.

Colors from Black and White (Pattern A)

This rotating black-and-white pattern stimulates the color receptor cells in your eye in a very peculiar way so that you see colors after staring at it for a while. This optical illusion might be related to the different response of the cones in our eyes. The colors vary across the disk because at different radial positions on the disk, the black arcs have different lengths, so the flashing rate they produce on the retina is also different. Reverse the direction of rotation and compare the order of colors you see. Also try varying the speed of rotation and compare the results with your initial observations. What happens if you change the thickness of the arcs? Does the illusion persist when you do that?

Spinning Spiral (Pattern B)

When you stare at a moving pattern, your eyes and brain get used to seeing movement. When you look at your hand standing still, your eyes see movement in the opposite direction. Reverse the direction of rotation of the disk. After staring at it for a while, look again at your hand. What do you see now? What would happen if you stared at two spinning spirals at the same time, with opposite directions of rotation?

White from a Rainbow Disk (Newton's Disk, Pattern C)

Our eyes can distinguish distinct images only if the time interval between two consecutive images is greater than a certain value. If a sequence of images is presented in a very fast way, as in a film, we cannot identify each image individually. So, when the disk of rainbow-colored sectors spins, the colors merge at the eye to produce white. The different response of the various cones in our eyes might explain the choice of colors in Newton's disk and the differences in size of the colored sectors. Change the order of the colors in Newton's disk or the size of the sectors and see if the new disk still produces white when you let it spin.

The Crazy Roulette Wheel (Pattern D)

It is just a bent version of the rows of light and dark tiles that appear wedge-shaped (see "Bending Parallel Lines" earlier in this experiment). What about inventing other patterns that also produce cool optical illusions? You can start by modifying some features of the above patterns and see if something new turns up.

4

MOIRÉ PATTERNS

Discover new images using a mirror and a screen or a comb.

SUPPLIES

- flat mirror (your bathroom mirror will work)
- silk, or wire or plastic screening with very small holes, or a colander or sieve, or even a comb
- PC (optional)

STEP BY STEP

Bring the screen very close to the mirror and watch what happens. Vary the distance between the screen and the mirror, and then try varying the distance between you and the screen. If you have a comb, repeat the experiment and watch what happens.

ALTERNATIVE

Fold a piece of the silk over and slide the top part slowly over the bottom part. What images can you see?

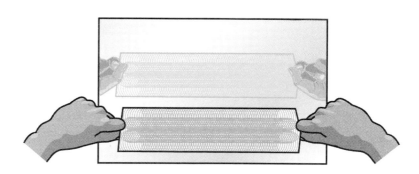

TWO STEPS FURTHER

A. Stretched Silk

Stretch a piece of silk (a shirt will do it) against your PC or TV screen and see what happens. How you can explain that?

B. Meshed Lines

Draw a mesh of lines on the screen of your PC with a drawing program. Now shrink the mesh with the zoom-out feature, gradually making it smaller. What happens? Why?

FUN FACTS

When you superpose two regular patterns slightly displaced from each other, in some areas the two patterns reinforce each other in your eyes, thus producing a contrast. Now, take an ordinary lens and examine the screen of your computer. (A white screen is ideal.) You can use a transparent plastic bottle filled with water as a lens. What do you see? Is there a subtle pattern hidden in the white screen? By stretching a silk against the screen, you can easily demonstrate that superposing two repetitive patterns greatly magnifies their differences. Does the same reasoning apply to the shrunken mesh drawn on your PC screen? Moiré patterns can be found everywhere. They are just waiting for you to discover them.

5

LENSES MADE OF AIR AND WATER

Make your own lenses and see the world through different eyes.

SUPPLIES

- transparent plastic bottle (cylindrical form) and screw-on lid, clean and with the label removed
- filament-type standard light bulb
- PVC pipe 3/4 in (2 cm) wide and 4 in (10 cm) long
- large cork
- glass of water
- superglue
- flashlight

A. Lens in a Bottle/Cylindrical lens (shape of a cylinder)

STEP BY STEP

Fill the bottle with water and close it tightly. Your lens is ready. Write on a piece of paper the words DODO and MAMA. What happens as you slowly pull the lens away from the paper? Now touch the lens to a TV screen or computer monitor. What can you see? In a dark room, shine the flashlight against a wall. Now place the flashlight against the lens and see what happens. Place a finger pointing upward behind a cylindrical glass of water and see what happens to your finger. Plunge the finger in the water at different distances from the glass wall and see what happens. Compare the Styrofoam ball or cork of your acceleration gauge (Experiment 36, "Accelerometer," under *Fun with Mechanics*) with another ball or cork outside the glass jar.

B. Spherical Lens ★★

CAUTION ————————————————
This experiment requires the help of adults.

STEP BY STEP

Wearing protective goggles, carefully remove the glass piece inside the bulb that holds up the filament. See illustration (1), and follow these steps:

1. With a craft knife, lift off the point of the bulb. Then use needle-nose pliers to take the whole metal bottom off.

2. With the points of the needle-nose pliers, break the glass piece that holds the filament. Widen the opening so that you can remove the entire filament with its glass support piece.

Hold the bulb with a thick cloth or glove to avoid accidents in case the bulb breaks. Heat the end of the PVC pipe with a flame (a candle will do it), turning the PVC constantly so its end heats evenly, as shown in illustration 2. Carefully push the metal end of the bulb all the way into the PVC pipe (illustration 3). Submerge the bulb and pipe into a basin of water so that both are filled up. Seal the PVC pipe with the cork (if the cork is too small, wrap it in masking tape for a better fit). Your spherical lens with handle is ready. To test it, focus sunlight on a piece of paper. Watch objects and people through the lens and pay attention to what happens as you increase the distance from your eye to the lens.

1

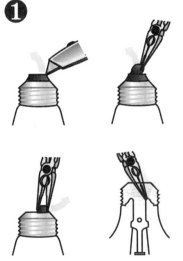

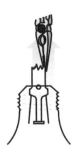

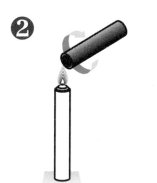

Alternative 1

Cut a plastic bottle to make a small funnel. Heat the end of the PVC pipe over a candle flame (illustration 4), turning constantly, and fit the pipe into the funnel, as shown. Next, cool the pipe with running water. With a craft knife, cut the excess plastic off the funnel, near the lip (see illustration). Fill the set (bulb and PVC) with water and screw on the bottle lid. If necessary, use superglue to seal the joints.

Alternative 2

Fill up a goblet and look at your finger or another object behind it. Try to focus the beam of a flashlight with the goblet near a wall. How about placing the goblet in the sunlight?

Alternative 3

Fill up a clear plastic bag and tie a tight knot in its neck. This is your lens. Repeat the experiments suggested above.

Alternative 4

Blow up an airtight plastic bag and tie its neck in a tight knot so the air doesn't escape. Submerge yourself, along with this new lens, in a swimming pool and look through this lens. What do you see? Swimming goggles have the same effect. If you are near-sighted, all the better!

FUN FACTS

Lenses are based on the fact that light "bends" when it crosses the interface separating two transparent regions where light propagates at different speeds, like air and glass or air and water (see Experiment 1, "Invisible Glass"). To make parallel rays coming from a distant source (the Sun or other star, for example) meet at a certain point (the *focal point* of the lens), the surface of the lens must be round. A flat surface would affect parallel rays the same way, so the rays would stay parallel and never meet. A cylindrical lens is a special case, since it combines roundness and flatness along the lens axis. How does that combination affect the image? Besides the roundness of the interface, the refraction indexes of the lens and its surroundings define the properties of the lens. Compare a transparent plastic bottle filled with water in air and an empty bottle (filled with air) immersed in water. Compare also your lens made of a light bulb filled with water and an empty bulb immersed in water (see also Experiment 15, "Exploring the Laser Ray," part G).

6

THE LIGHT AT THE END OF THE TUNNEL

There is light at the end of the tunnel! You just need to know how to see it.

SUPPLIES

- flat mirror with the edges sanded smooth
- 2 PVC pipes, 1 1/2 to 2 in (4–5 cm) in diameter
- wire
- hinge or piece of flexible plastic
- flashlight

STEP BY STEP

Attach the hinge or flexible plastic to both PVC tubes with the wire in a way that allows you to change the angle between them (see the figure). Place the tubes in a V-shape, with the flat mirror near to where they meet. Light the flashlight and place it solidly inside one of the tubes so it doesn't slip. Changing the angle between the tubes, watch what happens from the opening of the other tube.

HINT

Cover the mirror with a sheet of white paper and find out how wall paint affects room illumination (diffuse reflection).

A STEP FURTHER

The next time you play pool, take your project with you. Repeat the experiment by placing the mirror parallel to the border of the pool table. Then push the ball along the direction of the pipe with flashlight and mark the path of the ball after it hits the border. Compare the light path with that of the ball. Is there any difference? You can also do the experiment with a mirror and a laser pointer.

FUN FACTS

Each ray we see reflected by the mirror chooses the path that takes less time to travel from the light source (its origin) to our eye (its final destiny). This is the efficiency principle at work (see Experiment 1, "Invisible Glass"). It is easy to find the point where each ray hits the mirror. Since light in air propagates along straight lines (not always, though; see Experiment 3, "Bending Laser Beams with a Hair Dryer," under *The World of Atoms and Our World: Cold, Heat, and Giant Bubbles*), the total distance traveled by the ray is the sum of two line segments separating points S (source) and M (where the ray hits the mirror) and M and E (eye). First, draw point S' behind the mirror at the same distance as point S, so that SM equals S'M. The total distance traveled by the light ray can then be expressed in terms of S'M + ME. Since the shortest distance between points S' and E is along a straight line, point M is the intersection of the line segment S'E with the mirror. Can you now find in advance the angle between the left pipe and the mirror so that you can see "the light at the end of the tunnel"?

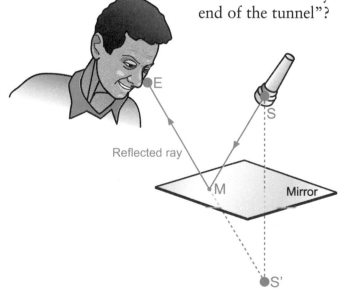

7

THE GHOST BEHIND THE MIRROR

Discover the ghosts that live behind mirrors.

SUPPLIES

- 2 clothespins
- flat mirror (6 × 6 in or 15 × 15 cm, for example) with the edges sanded smooth
- 2 pens or pencils (use two identical thin objects)

STEP BY STEP

Make sure the edges of the mirror are smooth so that no one will be cut accidentally. Place the mirror vertically upright, using, for example, two clothespins (see illustration). Place one object in front of the mirror and one behind, at the same distance, one in front of the other as shown. Stick the one object to the back of the mirror and compare it to the image formed. What can you conclude?

HINT

Change the position of the object behind the mirror and see what happens. Compare the object with its image. Have you ever noticed that objects seen in the rearview mirror seem further away than they really are?

ALTERNATIVE

Hold the mirror in front of you and look at the image formed. Place any kind of object behind the mirror (for example, a ruler, pencil, or book). See how easy it is to put something inside your head without having to open it. Why not try this experiment on your friends?

An illuminated object can be viewed as an infinite set of points emitting light in all directions. Since we cannot trace back the paths of the rays emitted by each point, we prefer to think that all light comes straight to our eyes. If all those rays that come to us are projected behind the mirror, they will all meet at a certain point (which one?). Now add the projected rays of all light sources and the ghost becomes visible.

8

LEVITATION AND CUBISM WITH A FLAT MIRROR

How to levitate and reinvent cubism using only a flat mirror.

SUPPLIES

• flat mirror with the edges sanded smooth

STEP BY STEP

Make sure the edges of the mirror are smooth so that no one will be cut accidentally. Hold the mirror straight up vertically. Have someone place the tip of his or her nose to the side of the mirror, so that his or her body is divided right down the center (as the picture shows). To "levitate," you need only to hold one foot on the ground behind the mirror and lift the other foot. To make a new "cubist" image, turn your head slightly toward the person looking in the mirror (as the other illustration shows). Even funnier will be putting one finger beside the person's head to produce a pair of horns.

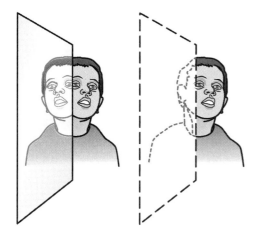

FUN FACTS

People usually have an almost symmetrical body. To check this, ask someone to stand with the edge of a mirror bisecting his or her body. The image you see replaces the hidden part of the body behind the mirror, so that you see a symmetrical whole. To duplicate the number of noses, just ask the person to move his or her head off the edge, as indicated. To fly, just straddle the mirror and raise one leg. The public sees only one true leg, levitating. The image produced by the mirror does the rest of the trick (see previous experiment) and levitation is accomplished, provided you can stand on only one foot. Some say that behind all magic is smoke and mirrors. In this case, it's just physics and fun!

9

MAGICAL THEATER

Build a magical theater to mesmerize children and adults alike.

SUPPLIES

- cardboard or wooden box, about 1 × 1 × 1.3 ft (30 cm × 30 cm × 40 cm)
- piece of glass, with the edges sanded smooth, measuring 0.92 × 1 1/4 ft (28 × 38 cm) for the box size indicated above
- 2 flashlights (with batteries)
- construction paper (black matte)
- colored construction paper (optional)
- 2 small figurines or other cool objects
- thicker cardboard or paneling (box lid)
- masking tape
- white paper glue or glue stick

STEP BY STEP

Make sure the edges of the glass are smooth so that no one will be cut accidentally. Make an opening (about 8 × 8 in or 20 × 20 cm) in one of the sides of the box for the stage, as shown on page 162. Line the inside of the box with the black construction paper. Make a lid for the box (1.05 × 1.35 ft or 32 × 41.2 cm) out of the thicker cardboard or paneling, with two holes in the places indicated in the picture. The two flashlights should fit tightly in these holes. Place the glass on its side as indicated. Seal it in place with the masking tape. Place the two figurines or other objects where the illustration indicates and place the top on the box, with the flashlights in their positions. The show should take place in a dark room. Turn on the flashlights one at a time and check out the stage. If necessary, adjust the position of the glass, dolls, or box lid so that the actors are well lit. To make your little theater more exciting, cover the outside of the box with colored construction paper and decorate it. Your magical theater will be smashing! It will enable you to perform several cool tricks. For example, you can levitate objects—just place them on a small black matte support at the corner close to the stage. Feel free to invent your own performances for the public's delight!

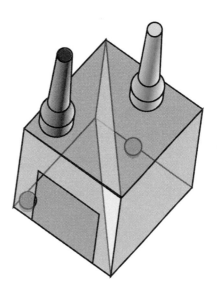

FUN FACTS

Every transparent surface can both transmit and reflect light. You probably have already noticed that a transparent glass window, which allows you to see through it, changes at night when the outside is less illuminated than the indoors. What dominates in daylight, and what dominates at night? Looking at the positions of the objects in the box, discover where, in the public's eyes, the object reflected by the glass seems to be located. (See also Experiment 6, "The Light at the End of the Tunnel.") Are you now ready to explain for your public how the magical theater works?

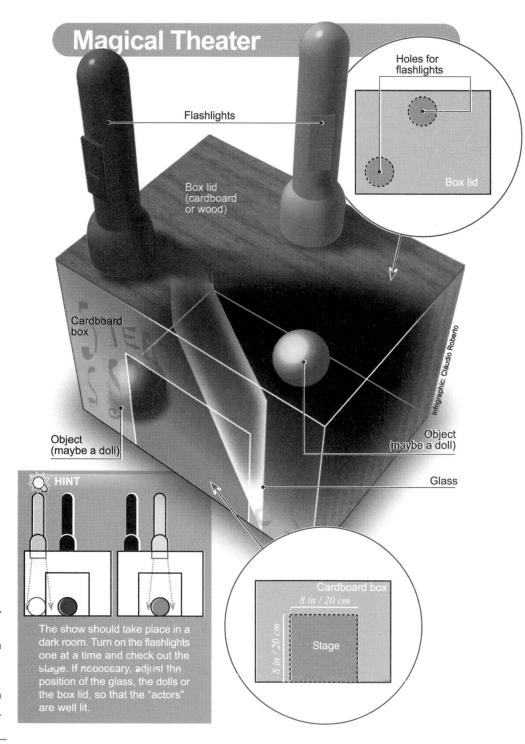

Magical Theater

Flashlights

Box lid (cardboard or wood)

Holes for flashlights

Box lid

Cardboard box

Object (maybe a doll)

Object (maybe a doll)

Glass

Infographic: Cláudio Roberto

HINT

The show should take place in a dark room. Turn on the flashlights one at a time and check out the stage. If necessary, adjust the position of the glass, the dolls or the box lid, so that the "actors" are well lit.

Cardboard box

8 in / 20 cm

8 in / 20 cm

Stage

10

THE MIRACLE OF THE FISHES: PARALLEL MIRRORS

We can all make miracles come true!

SUPPLIES

- 4 clothespins
- 2 identical flat mirrors (6 × 6 in or 15 × 15 cm, for example) with the edges sanded smooth
- craft knife
- steel wool pad
- tiny plastic fish (or any small object)
- clear glass panel in the same or similar dimensions as the mirrors

STEP BY STEP

Make sure the edges of the mirror are smooth so that no one will be cut accidentally. Place one of the mirrors on a soft cloth or towel so as not to scratch it. Using the knife, scrape a bit of the surface off the back of the mirror to make a small opening, around 1/4 in (5–8 mm) in diameter. Use the steel wool to smooth out the edges of the opening. Place the mirrors vertically upright, facing each other, held up by the clothespins, as shown. Place the tiny fish or other handy cool object between the mirrors. Through the opening, count the number of images you can see.

Replace the mirror you looked through with a glass panel. In a dark place, light up the glass with a flashlight and see what happens. If necessary, cover the sides and the top of the space between the mirrors with a black cloth.

The object between the two mirrors is reflected by both of them. The images formed, in their turn, are also reflected. So, you end up with an infinite number of images (the image of the image of the image, etc.). Why do the reflected images seem to become fainter as their distance to the mirrors increases? (Remember, they are all "ghosts"!) Do the mirrors absorb some of the light? You can try conventional (or back-surface) mirrors (with the silvering on the back of the glass) and then *first-* or *front-surface mirrors* (with the silvering on the front of the glass).

11 KALEIDOSCOPES FESTIVAL

Take part in this unique festival. Transform an ordinary experience into something extraordinary. Discover the fascinating world of kaleidoscopes and find out how they work.

SUPPLIES

- 3 pieces of mirrors, about 1 1/2 × 6 in (4 cm × 15 cm), with the edges sanded smooth, or 6 pieces of a CD, 1 × 2 3/4 in (2.5 × 7 cm)
- adhesive tape
- sheet of white paper
- tiny pieces of confetti or colorful objects
- black matte paper

HINT

To build your kaleidoscopes, you can use conventional (or back-surface) flat mirrors (with the silvering on the back of the glass). These mirrors might create duller, more blurred reflections (see Experiment 10, "The Miracle of the Fishes"). The CD model suggested here is more appropriate for children to lessen the chance of injury in case of breakage. (You can also use Plexiglas mirrors for this purpose.) The best mirrors to use, however, are first- or front-surface mirrors (with the silvering on the front of the glass), which give sharp, clear reflections.

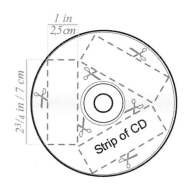

1 in
2,5 cm

2 3/4 in / 7 cm

Strip of CD

A. Triangle Kaleidoscope

STEP BY STEP

Make sure the edges of the mirrors are smooth, so that no one will be cut accidentally. Tape the edges of the mirrors together, making a closed triangle, with the reflective surfaces on the inside. Look through the closed triangle. Is it fun? Seal one end of the triangle shut with the white paper and tape. Place tiny pieces of confetti or colorful objects inside. The second end should be closed with a piece of white paper with an eyehole in it. Your kaleidoscope is ready! Look through the eyehole while turning the kaleidoscope. What happens if you cover one of the mirrors with a strip of black matte paper?

> **HINT**
>
> If you use a CD, carefully score the back of the CD with a craft knife, then just bend it—it should break at the right spot easily. To make a long enough kaleidoscope, make double the number of pieces and join them together with masking tape, as the illustration shows.

B. Odd Kaleidoscope

Build a triangular kaleidoscope with flat mirrors, but instead of a triangle with equal angles, use angles of, for example, 30, 60, and 90 degrees between the mirrors to combine different symmetries. To obtain an "odd kaleidoscope," use pieces of mirrors with different widths, such as 1 3/16, 1 5/8, and 2 in (3, 4, and 5 cm), but all with the same length (6 in/15 cm) to see what happens when symmetry conditions are not satisfied. (You can

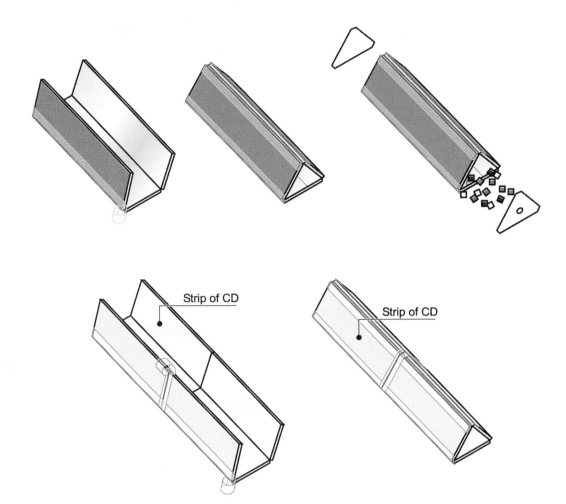

Strip of CD

Strip of CD

build kaleidoscopes with triangular cross sections with arbitrary angles between mirrors, provided their sum is 180 degrees.) Compare the images produced by a "normal" kaleidoscope and an "odd" one.

C. Fedorov's Kaleidoscope

STEP BY STEP

Follow the pattern shown in the figure to make a stand for the mirrors. Attach mirror 2 to the corner (base) of the stand using the masking tape. Join mirrors 1 and 3 with strong tape (electrical tape) so that the angle between them can be adjusted between 180 and 0 degrees (see the picture). Cover the inside edges of mirror1 with the masking tape so that it can move without scratching the mirror attached to the stand (base). Glue mirror 3 onto the back wall of the stand. Hold your finger (or a pen or pencil) over the corner of the two standing mirrors (1 and 3) and change the angle between them. What happens to the images formed in mirror 2? Now place your finger parallel to mirror 2 and bend it up and down. When you pull it further away from where the two mirrors meet, do the images grow? What relationship can you see between this kaleidoscope and the previous one?

A STEP FURTHER

Find out how night retro-reflectors (corner reflectors, or "cats' eyes"), used on road surfaces, work when the headlights of a car illuminate them. With mirrors 1 and 3 making a 90-degree angle, shine a flashlight on either of them and see how the reflected light is directed toward the car and its driver, and not wasted by going in all directions as with diffuse reflection.

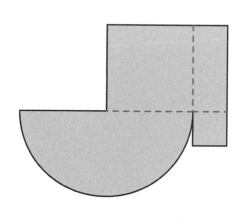

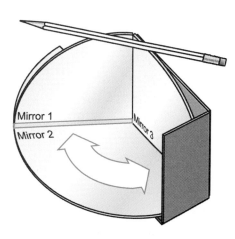

Mirror 1

Mirror 2

Mirror 3

SUPPLIES

- 4 identical flat mirrors in the shape shown in the figure, all of them with the edges sanded smooth
- electrical tape

D. Four-Sided Kaleidoscope

STEP BY STEP

Put the 4 mirrors side by side with their back sides up and apply tape to the joint edges. Leave a small separation between the edges, so that you can fold your kaleidoscope, changing its square cross section into a diamond-shaped figure.

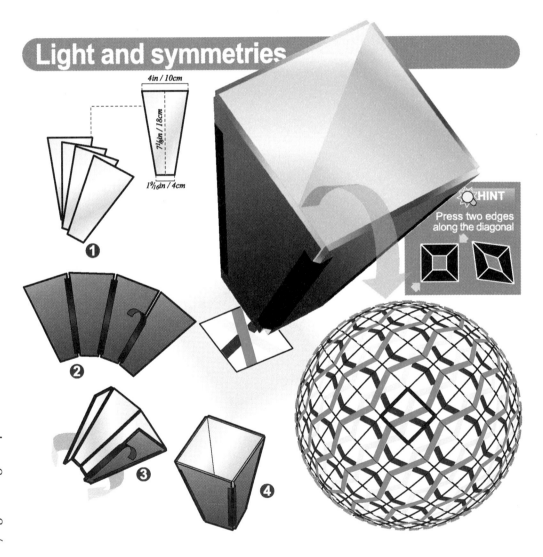

Light and symmetries

4in / 10cm

7⅛in / 18cm

1⁹⁄₁₆in / 4cm

❶

❷

❸

❹

HINT

Press two edges along the diagonal

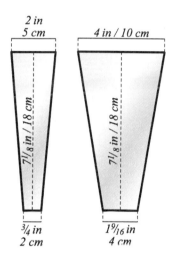

2 in
5 cm

4 in / 10 cm

7¹/₈ in / 18 cm

7¹/₈ in / 18 cm

¾ in
2 cm

1⁹/₁₆ in
4 cm

A NEW VARIANT: EXPLORING TWO PAIRS OF MIRRORS IN DIFFERENT SHAPES

STEP BY STEP

Make sure the edges of the mirrors are smooth so that no one will be cut accidentally. Put the four mirrors side by side, alternating the smaller and larger pieces with their back sides up, and apply electrical tape to the join edges, as you did in model C, Fedorov's Kaleidoscope. Leave a small separation between the edges so that you can fold your kaleidoscope, changing its rectangular cross section into a diamond-shaped figure.

How to Use Model D

In an illuminated room, place a flat figure, preferably colorful, on a working table, or choose a poster or picture hanging on the wall. Keep the smaller opening (base) on the figure or close to it and look at the image through the larger opening. Explore the infinite possibilities of your new kaleidoscope by scanning the different parts of the figure. In each position of the base, fold the kaleidoscope by gently squeezing two opposite edges. Be prepared to spend hours on end looking around with this fascinating optical gadget! You will discover worlds within worlds. Put your kaleidoscope on a reduced map of your area and you will see it projected on the world's map! If you are a fan of American football, the new variant will bring you unforgettable surprises!

Playing with Light: Optics

- 1 pair of rectangular flat mirrors and 1 pair of flat mirrors in the shape shown in the figure, all of them with the edges sanded smooth
- electrical tape
- cardboard

E. Moving Mirrors Kaleidoscope

STEP BY STEP

Make two frames for the larger mirrors out of the cardboard, as shown. Make sure that the space between the two parallel mirrors is slightly larger than the width of the rectangular strips of mirror. Attach the edge of a cardboard rectangle (lateral walls) to the back of the mirror strips using electrical tape, as indicated. Fix the rectangles with mirrors in the two frames (use electrical tape to keep the cardboard pieces fastened together).

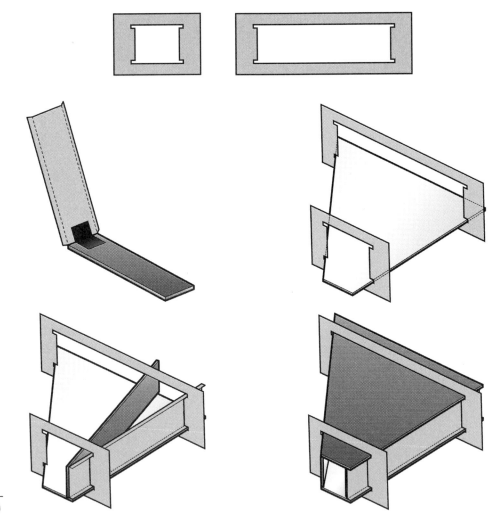

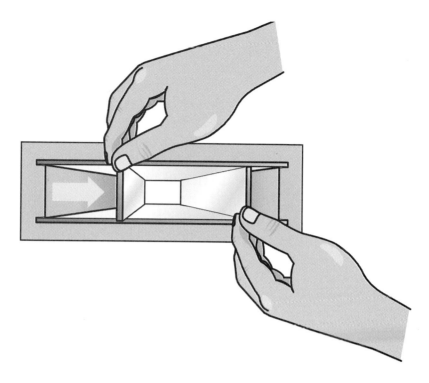

How to Use Model E

In an illuminated room, put a flat figure close to the smaller opening of your kaleidoscope and observe the image through the larger opening. Move the strips of mirror back and forth, and enjoy the fanciful images that this gadget allows you to see.

FUN FACTS

If you closely examine the images obtained with kaleidoscopes, you will realize that they are simply a result of multiple reflections and a special arrangement of mirrors. The functioning of a kaleidoscope is basically determined by the angle between the flat mirrors and the number of mirrors used in the kaleidoscope. Put two flat mirrors close to each other with an object in between. Vary the angle between the mirrors and the images will multiply (you can use model C to do this). When you add extra mirrors to obtain a kaleidoscope, the images are further reflected, forming patterns with beautiful symmetries. You can trace back the image formation in a kaleidoscope by considering a single-point light source (S) inside the kaleidoscope. The image of S as seen by an observer can be obtained by considering a point light source S' behind each of the mirrors at the same distance as S (see Experiment 6, "The Light at the End of the Tunnel" in this part of the book). Now, you just have to find the multiple images of S' produced by the kaleidoscope's mirrors. In this way, you can discover why in "odd mirrors" the multiple images do not form a closed loop of point light sources for arbitrary angles between the mirrors. Using this simple trick, it is easy to find out how you get three-dimensional images with pairs of tapered flat mirrors in models D and E. (You can explore with model E the images formed by the two parallel mirrors and also replace them with pieces of black matte paper—what happens then?) Simple geometry allows you to demonstrate that the rays reflected by perpendicular flat mirrors are parallel to the incoming rays. (Check it out with model C. Use a pocket flashlight. Do not use a laser beam, because it can reach your eyes and damage your retina.) Could you use curved mirrors to build kaleidoscopes? With so many possibilities, are you ready to invent your own model?

DARK CHAMBER

Experience a new perception of the world with a unique dark chamber.

SUPPLIES

- black matte posterboard
- tracing paper
- paper glue
- standard filament light bulb (a clear glass bulb gives more interesting results), 100 to 150 watts with socket and connection

STEP BY STEP

Make two boxes out of the posterboard, following the two figures drawn here. One should fit loosely inside the other, with about 1/8 in (3 mm) of space between them. The dotted lines in the illustration show where the posterboard should be folded to make a box, and the solid lines show where it should be cut. Make a slot about 1/16 in (1.6 mm) wide and 2 in (5 cm) long on the bottom of each box, as shown. Also, make a posterboard frame for the tracing paper ("screen"), as the drawing shows. In the smaller box, make slots A, B, C, and D for the frame to fit into. To make the standard dark chamber, push the smaller box all the way to the back of the larger box until the ends touch, with the slots perpendicular to each other. Point your dark chamber at the filament with the lamp

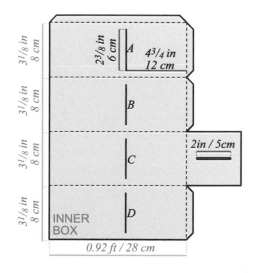

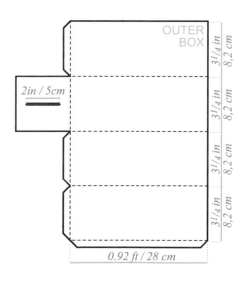

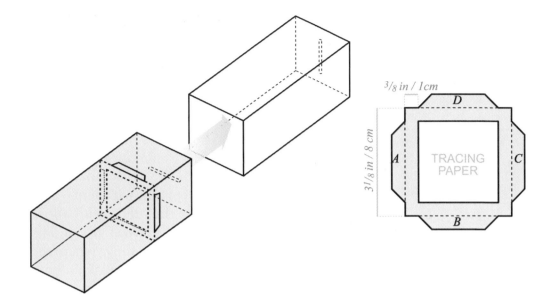

D

A   TRACING
      PAPER   C

B

switched on. Watch the images formed on the screen as you gradually increase the distance between the two boxes by sliding the smaller box slowly out. You can also try your dark chamber outdoors by pointing it at a well-lit object. Wait for a while until your eye becomes used to low illumination. Do you see color at all?

A STEP FURTHER

Take the smaller box and screen out of the larger box. Slowly bring the clear light bulb closer to the slit in the small box with the light switched on. What do you see on the screen? Use your fingers to produce a "pin-hole" at different positions in the slit. What do you see now? What do you conclude about the image formed by the slit as whole?

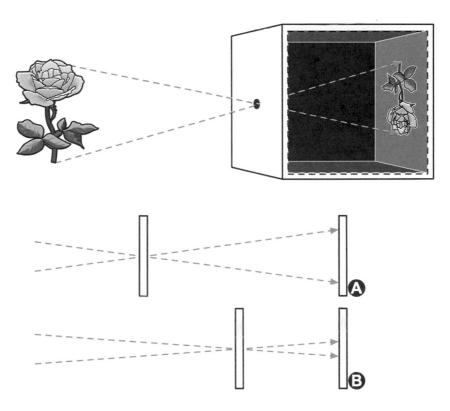

FUN FACTS

The principle of a standard dark chamber (ends of boxes touching) is illustrated in the following figure. Light rays emitted by an object cross through the pinhole and form an inverted image on the screen. The greater the distance between the screen and the pinhole, the larger the image becomes. In this case the pinhole projects points on horizontal and vertical lines with the same magnification for both directions. When the slots are separated, their distances from the screen become different, so the magnifications they produce are also different (see figures A and B). Each slot projects only lines perpendicular to the slot. This means that the magnifications for the horizontal and vertical directions are different, which creates the distortion observed in the image. Now, what happens if the slots are curved?

NEW DISCOVERIES WITH POLARIZERS

Discover what polarized light and double refraction are.

SUPPLIES

- 2 Polarizers (polarizing filters), found in photo shops (some sunglasses are also Polarizers)
- cover of a CD case or a piece of hard plastic
- laser pointer
- clear glass
- 2 pieces of clear adhesive tape
- 1 folded and crumpled sheet of cellophane
- 1 ordinary flat mirror

STEP BY STEP: POLARIZED LIGHT

Try looking through a polarizer at the light reflected by a CD case. (Hint: Pay special attention to the corners.) Rotate the CD case, then tilt it or look at it from a different angle. What happens? Now, look at the CD case in reflection, using an ordinary mirror. Do you notice anything special? Substitute a shiny piece of black plastic for the mirror. Compare the image formed by reflection and the object that is reflected. Is there anything weird in there? Turning the polarizer in front of your eye, look at the blue sky and a rainbow. (See Experiment 14, "Why Is the Sky Blue?") Look, too, in different positions at the reflections on a wall or window and find out why polarizers are used for cameras. Discover also why some fishermen use glasses with polarizing filters to better spot fish through the water. In a dark room, hold the polarizer against the laser end of a laser pointer pointed at a wall and see what happens to the laser light that hits the wall when you turn the polarizer. So many fascinating mysteries!

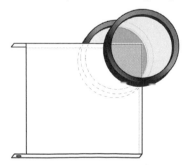

A STEP FURTHER: BIREFRIN-GENCE/DOUBLE REFRACTION

Look through both polarizers placed over a light source. Turn one of them until they both look dark (crossed polarizers). Place the plastic (CD case top) between the two polarizers, as the illustration shows, and see how the plastic is transformed. Tilt the plastic or rotate it and see what happens. Use a folded and crumpled piece of cellophane and repeat the previous experiment.

Stretch one piece of the adhesive tape before taping it down on the glass. Tape the other one down beside the first without stretching it. Now, place the glass between the two crossed polarizers and compare the two pieces.

FUN FACTS: POLARIZED WAVES

Light can be described in terms of transverse waves. You can produce transverse waves with a stretched rope, for example, by fixing one of its ends and moving the other end either up and down or sideways *(linearly polarized waves)*, or moving it around in a circle, either clockwise or counterclockwise *(circularly polarized light)*, as indicated in the figure. The oscillations you produce are perpendicular to the direction of propagation—hence the name transverse waves. Waves are characterized in general by a length (the wavelength), which measures how often a wave pattern repeats itself in space when you take a snapshot of the whole wave, and by a frequency, which measures how often the wave repeats itself in time when you consider a fixed point in space along the wave's trajectory.

FUN FACTS: POLARIZERS

A polarizer works as a filter of oscillations. You can simulate a polarizer with two parallel broom handles close to each other. First, you produce a wave in a stretched rope (see figure). Then, you place the broom handles at a point where the transverse oscillation is greatest (an "antinode"). If you place them at a point where the cord is motionless (a node), the trick will not work! The "polarizer" lets pass through only oscillations along a certain direction (the polarizer axis) and eliminates the remaining oscillations. A circular polarized wave, for example, is transformed into a linearly polarized wave as it crosses a polarizer (see figure).

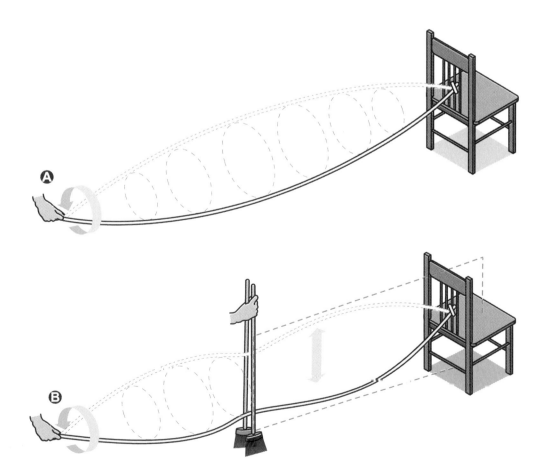

FUN FACTS: LIGHT WAVES

Light waves consist of oscillating transverse electric and magnetic fields. An electric field accelerates a charged particle and a magnetic field exerts a force on a moving charged particle when its velocity has a component perpendicular to the magnetic field. The light from the Sun and the light emitted by ordinary lamps is a mixture of waves with different wavelengths oscillating in all possible directions on a plane perpendicular to the direction of propagation (it is then unpolarized light). A real polarizer just lets the components parallel to its axis pass through and eliminates all other components. The polarizer thus produces linearly polarized light.

FUN FACTS: POLARIZED WAVES BY REFLECTION AND REFRACTION

When light strikes a surface, the reflection of light by the surface also produces linearly polarized light, especially at shallow incidence angles. The electrical charges on the surface act as tiny antennas. They respond differently to the oscillating electric field of an incoming light wave depending on its direction of oscillation. If the oscillation is parallel to the surface, the charges on the surface are pushed back and forth easily along the surface. In this case, the tiny antennas are very efficient and emit light linearly polarized along the horizontal direction. However, an electric field oscillating vertically pushes the charges up and down along the vertical direction. Since the charges can't leave

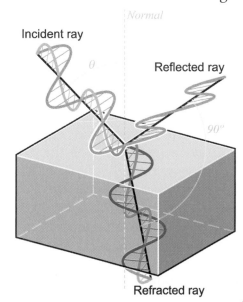

Incident ray

Normal

Reflected ray

θ

90°

Refracted ray

the surface, their response is very poor at shallow angles of incidence. In transparent surfaces, as the incidence angle is equal or greater than a particular angle (the so-called Brewster angle), the emission of vertically polarized light is completely cut off. There is another way to see how light polarization is produced by reflection. At the Brewster angle, the electric field of the transmitted (refracted) light oscillates vertically along the propagation direction of the reflected wave. The tiny antennas cannot generate a reflected ray with vertical polarization, since the corresponding electric field oscillation would be along the direction of the wave propagation. Because the incident light has both polarizations, the reflected light becomes horizontally polarized. The same mechanism produces polarized light in rainbows (see figure showing a glass filled with water simulating a rainbow; a laser beam can be used to probe the polarization effect of the rainbow—be careful with the extra light rays coming out of the glass!). Experiment 14, "Why Is the Sky Blue," also demonstrates how partially polarized light is produced in a blue sky. Sunglasses and the polarizer of photographic cameras can thus dramatically reduce glare by absorbing most of the light reflected from a surface. (To accomplish that, what should the orientation of the polarizer's axis be?) Light reflected from a highway surface at Brewster's angle is often annoying to drivers and can be demonstrated quite easily by viewing the distant part of a highway, the reflections of the Sun's light from car windshields, or the surface of a swimming pool on a hot, sunny day. Modern lasers often take advantage of Brewster's angle to produce linearly polarized light from reflections at the mirrored surfaces positioned near the ends of the laser cavity.

Now, considering that the Brewster angle varies with the frequency of the light components, can you understand what happens when you look at a CD case with a polarizer? Does a mirror produce polarized light? (Illuminate the mirror with a flashlight at different angles of incidence and check the reflected light with a polarizer.)

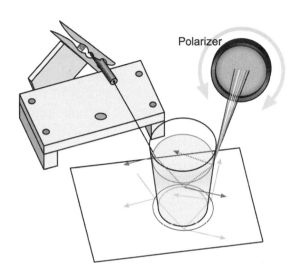

Polarizer

Simulating the Polarization Effect in a Rainbow ★

In a dark room, turn the polarizer (sheet of polaroid) to check whether the "rainbow ray" (laser beam coming out of the glass) is polarized. (It should be projected on a "screen," such as a white wall.) The dots represent the vertical polarization (perpendicular to the plane of the paper), and the double arrows stand for the horizontal polarization of the light's electric field.

CAUTION ————————————

To avoid accidents, place opaque obstacles at the spots where the reflected and the first refracted rays come out of the glass.

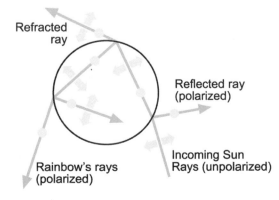

Refracted ray

Reflected ray (polarized)

Rainbow's rays (polarized)

Incoming Sun Rays (unpolarized)

FUN FACTS: BIREFRINGENCE (DOUBLE REFRACTION)

Birefringence occurs when a material displays two different refraction indexes for the horizontal and vertical polarizations of light. A light component with a polarization that affects the material most strongly has a higher refraction index and therefore travels slower than the light with the other polarization, associated with a lower refraction index. As a plastic is submitted to stress, it becomes *birefringent*. This can be demonstrated by using two crossed polarizers (sheets of polaroid). As unpolarized light crosses the first polarizer, it becomes linearly polarized along the polarizer's axis. Since the axes of both polarizers are perpendicular to each other, no light comes out from the second polarizer. If you place a piece of strained plastic in between the crossed polarizers, the polarized light coming out of the first polarizer is split into two components, one along the "fast lane" direction and the other along the "slow lane" direction. As the two wave components leave the plastic, there are several possibilities: the resulting wave coming out of the strained plastic can be linearly polarized, circularly polarized, and so on. (You can visualize the result of two perpendicular oscillations using a laser pointer, as suggested in Experiment 12, "Pictures of Sounds," under Playing with Sounds: Acoustics.) The several components of light are thus "filtered" by the second polarizer in different ways depending on how the plastic is strained and also by its thickness. Because the birefringence is related to the stress and is color-dependent, differently strained areas appear as different colors. If you bend a clear plastic ruler, does it display birefringence where it is bent? What happens when the ruler returns to its normal condition? Engineers use this trick to map out the strain distribution in plastic models of structures, such as bridges, submitted to variable loads. What about a CD case? When you glue pieces of rigid plastic or heat them up, do you produce birefringence?

Re-create the sun and the sky to find out why the sky is blue on a clear sunny day.

SUPPLIES

- clear plastic box or clear glass baking dish
- flashlight
- some drops of milk, preferably skim or nonfat
- Polarizer (optional)

STEP BY STEP

Fill the plastic box or glass baking dish with water. In a dark room, place it near a white wall. (This is your white screen.) Turn on the flashlight and light up the container, as shown. Check if there is light coming out of the sides of the container. Mix just a few drops of milk into the water, stirring it until well mixed. (Also, try adding dishwashing detergent instead of milk.) Shine the light on the container and check if light comes out the sides this time. What is the predominant color? If you have a polarizer available, use it to find out the nature of the light coming from the flashlight and the light coming out the sides of the container. Turn the polarizer at several angles along the sides of the container (see Experiment 13, "New Discoveries with Polarizers"). Compare the light reflected by the wall with the light coming out of the flashlight. Add a few more drops of milk and see what happens. At what times does the Sun's light travel the longest paths as it passes through the Earth's atmosphere: early morning, twelve noon, or at sunset? Imagine that the flashlight represents the Sun and the water with milk drops, a cloudless sky. What relationship can you find between this experiment and the color blue for the sky, or the fact that the sun often looks red at sunrise and sunset?

Playing with Light: Optics

When sunlight hits the Earth's atmosphere, it is scattered along all directions by atoms, molecules, ions, and electrons, which you can consider as very tiny balls. This phenomenon is called Rayleigh scattering. Imagine a high-pressure jet of water scattered by stones thrown in the air across the jet. The light scattering in the sky is different for each color. Red is far less scattered than blue. The color most scattered is violet. However, there is more blue than violet in the light coming from the Sun. Also, our eyes are much more sensitive to blue than to violet light. That's why we perceive the sky as predominantly blue. When most of the blue is scattered, red becomes dominant. The molecules of milk in the experiment act like tiny antennas. The scattered light you see from the side is emitted by the tiny antennas as a response to the oscillating electric field of the incoming unpolarized light wave of the flashlight (the "Sun"). The component of the incoming wave's electric field parallel to your direction of sight therefore does not show up in the scattered light if you look straight at the side of the box. The antennas cannot emit light in your direction of sight when they oscillate along this direction, so the scattered light you see is polarized. (Can you use this scattered light to map out the stress in a clear plastic box?) When you look at different angles, the scattered view changes a bit, since now your sight direction no longer coincides with any of the directions of oscillation of the tiny antennas. Can you understand now why the blue light of the sky is partially polarized? Can you also find out along which sight direction the light is most polarized?

Transform a laser pointer into a source of discovery.

SUPPLIES

- laser pointer
- wooden holder for the laser pointer (see experiment E, "Foil Reflections," in this sequence)
- clear plastic container or glass baking dish (no lid)
- water
- a few drops of milk, preferably skim or nonfat
- triangular kaleidoscope (made in Experiment 11, "Kaleidoscope Festival")
- a few sheets of white paper
- roll of aluminum foil
- piece of silk (silk clothing, for example)
- standard filament light bulb (clear glass bulb)
- clear plastic tube from a pen, cylindrical or sextagonal, without the ink cartridge
- flat wooden square
- a clothespin
- Polarizer (optional)

CAUTION

Never point a laser light at your own eyes or at the eyes of another person. A laser light can cause permanent damage to the retina. Also, take care not to point it at a mirror in a way that it could reflect into anyone's eyes. Following these safety procedures, you can have a great time exploring new universes.

These experiments should be done in a dark room. All of them, except F, "Lens Made of a Drop of Water," and H, "Silky Laser Beam," use the same container with the mixture of water and a few drops (a very few drops!) of milk.

STEP BY STEP

A. Through the Walls

Point the laser toward one of the walls of the container. Can you "see" light coming out of the other side of the container (forward ray) or off the water surface (scattered light)? (See Experiment 14, "Why Is the Sky Blue?") If you have a polarizer available, use it to examine the light coming out of the side and the top. Hint: Turn the polarizer while you look through it.

A STEP FURTHER

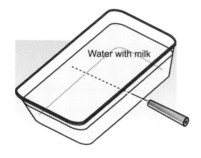

Start with the container empty and the laser held fixed, as shown in the figure (use a support to keep the laser fixed; see experiment E in this sequence). At the spot where the laser beam hits a wall, place a mirror almost upright, so that the laser is projected back onto another wall or on the ceiling (your screen). Now, you can see the effect of the container's walls alone on the laser beam. Next, fill the container with water and add just few droplets of milk to it. (You can later add extra droplets to see what happens.) Mix the water and wait until it comes to rest. Then follow up the changing pattern projected on the screen superposed on the fixed pattern you see when the container is empty.

FUN FACTS

The water molecules wander around inside the container due to their thermal energy. The milk molecules (mainly proteins, lactose, and fat globules—the biggest of all of them), which by comparison are very few, are constantly struck by the much smaller water molecules and displaced randomly (this is called the Brownian motion). From time to time, milk molecules are kicked into the path of the laser beam. They scatter the light, producing wavelets. These wavelets overlap and interfere, producing the dynamic pattern you see on the screen superposed on the fixed pattern produced by the reflections of the beam in the walls of the container. The laser beam then allows you to follow up the subtle dance of the invisible milk molecules. The farther the mirror is from the screen, the bigger the visual effect produced. Now, you can replace the container with a clear glass filled with hot water (be careful not to spill it!). Discover how convection currents associated with hot water moving up and cold water moving down affect the laser beam. Welcome to the jiggling nanoworld!

B. Covered Container

Cover the container with the white sheet of paper. Tilt the laser to point it from the surface side of the container, as shown, varying the angle that it enters the water (incidence

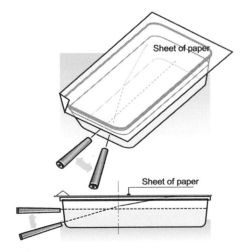

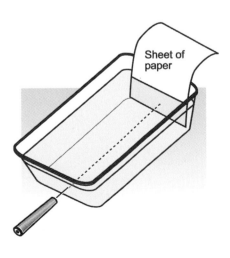

angle) until the beam is almost touching the water surface. See if light touches the sheet of paper as you tilt the laser beam toward and away from the surface of the water. Now, uncover the container and submerge a piece of white paper into the water, keeping it touching the sides. Looking from above, point the laser toward the paper-covered side of the container and watch the light that hits the paper. Take the paper off and shine the laser the same way again. Compare what you saw with the paper and without. Does the white paper produce diffuse reflection? (See the hint in Experiment 6, "The Light at the End of the Tunnel.")

C. Behind the Mirror

Place a mirror next to one of the sides of the container. Shine the laser into the side opposite the mirror so that the beam is first perpendicular to the mirror, and then tilted or angled. Looking from above, see what happens to the reflected beam of light. Could it be the "Ghost Behind the Mirror" (Experiment 7 in this part of the book) appearing again?

D. Kaleidoscopic Image

Take the paper ends off the triangular kaleidoscope that you made in Experiment 11, "Kaleidoscope Festival." Make a small opening at the back of one of the mirrors, as explained in Experiment 10, "The Miracle of the Fishes." The opening should allow the laser beam to pass through. Submerge the end of the kaleidoscope in the water so that the opening is a bit below the surface of the water and touching one of the sides of the container. Point the laser beam through the side of the container toward the opening

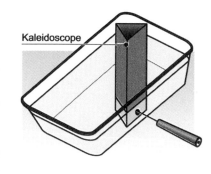
Kaleidoscope

and watch what happens from above. Try different angles of incidence. How about lighting up one of the corners of the outside of the kaleidoscope by pointing the laser beam toward the bottom of the container?

E. Foil Reflections

Fix the plastic pen tube upright in a hole in the wooden support, as shown in the figure (see also page 181), so that it intercepts the laser beam. The clothespin is used to keep the laser pointer turned on. The tube will act like a "cylindrical lens" (see Experiment 5, "Lenses Made of Air and Water"). Cut a piece of aluminum foil and place the more reflective side in a curved-over position, as the illustration shows, forming a concave mirror. Submerge it partway into the water, standing on its edge. Place the end of the laser pointer so it touches the tube, as shown, and see what happens to the beams reflected by

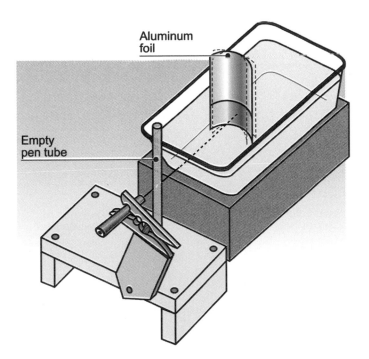

Aluminum foil

Empty pen tube

the aluminum foil. Try changing the shape of this convex mirror, as indicated by the dotted lines, and see what happens to the reflected beams. Make another mirror (convex), using the less reflective side of a piece of aluminum foil, leaning it up against the roll. Repeat the laser test.

F. Lens Made of a Drop of Water

STEP BY STEP

ADDITIONAL SUPPLIES

- 1-pint (500 ml) plastic bottle and screw-on cap
- water mixed with a few milk droplets

Make a small hole in the side wall of the bottle at the same level as the laser pointer fixed in the holder. Fill the bottle with water mixed with a few droplets of milk (to avoid the water escaping, hold your finger over the hole), then close the top tightly. Take your finger off the hole. In a dark room, place the laser pointer attached to holder on a work table near to a white wall (this is your white screen, where the laser beam hits). Place the bottle on the table with the hole in front of the laser pointer. Now, unscrew the top a little, just enough to form a small droplet in the hole, and then screw it on tightly. Switch the laser pointer on and follow the path of the light coming out of the droplet. Also have a look at the image formed on the white wall. Gently squeeze the side wall of the bottle, as

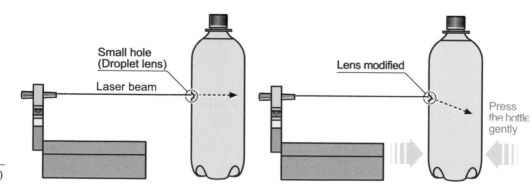

Small hole (Droplet lens)

Laser beam

Lens modified

Press the bottle gently

indicated, and see what happens. (Instead of unscrewing the top, you can simply start gently squeezing the bottle.) Can you control the focus of your new lens?

FUN FACTS

Since pressure propagates in water in all directions, the droplet is also affected when you squeeze the bottle. The droplet's curvature and the way it focuses light can thus be changed. This is a simple example of what is called *adaptive optics*. You can also use this model to illustrate how the human eye works and simulate some of its diseases, like myopia (short-sightedness) and astigmatism, related to deformations of the eyeball.

G. Submerged Light Bulb

Submerge the light bulb in the water. Place the laser pointer against the tube and watch the light coming out of the other side of the bulb. Compare this "lens" to the lens made with the bulb filled with water (Experiment 5, "Lenses Made of Air and Water"). Can you see how different they are?

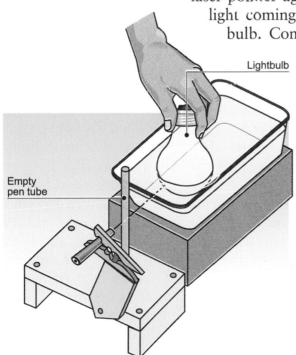

Lightbulb

Empty pen tube

H. Silky Laser Beam

Point the laser beam toward a wall. What happens to the lighted point as you move away from the wall? Cover the end of the pointer with a piece of silk and see what image is formed on the wall.

FUN FACTS

The end of a laser pointer is like the exit of a football stadium at the end of the championship game. As the light comes out, the beam opens sideways, just as the fans come out at the end of the game. A hole then diffracts light in the same way a complementary obstacle would do (the exit is equivalent to the stadium's gate in that the fans would avoid the gate if it were the only obstacle in their way). The many threads of silk also act as little "doors" for the laser light passing through. Place a strand of human hair on the end of the laser pointer and shine it on the wall in a dark room. You can also check what happens when you replace the hair with a narrow slit. If you weave together several strands of hair, you have the piece of "silk." It is easy to see why a slit (exit) and the hair (gate) scatter light the same way. If you cover a narrow slit with a hair, it will become opaque to light. That means that the light's electric field behind the covered slit is zero. You can think of it as the sum of the contribution from the slit and the hair considered separately. So, for a covered slit, the electric field of the light wave behind the slit should be the opposite of the electric field at the same position as you consider the hair alone. Since what we see has to do with the energy carried by light, it doesn't matter if the electric field is pointed one way or another.

TUBES OF LIGHT: FIBER OPTICS

Discover the principle of fiber optics.

SUPPLIES

- flashlight
- cardboard milk or juice carton
- piece of a drinking straw 3/4 in (2 cm) long
- superglue

STEP BY STEP

Make a hole in the bottom of the carton, about 1 in (2.5 cm) from the bottom, with a diameter a bit smaller than the piece of drinking straw. Put the straw 1/4 in (0.6 cm) into the hole and use superglue to seal around it. Make an opening in the top of the carton, just big enough to fit the flashlight. Cover the end of the straw bit with your finger and fill up the carton with water. In a dark room, place the lighted flashlight in the opening you made and take your finger off the hole. Let the stream of water fall onto the palm of your hand. You are demonstrating the basic principle of fiber optics (tubes of light), a technology that is replacing conventional copper

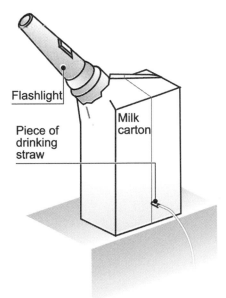

Flashlight

Piece of drinking straw

Milk carton

wiring for telephones lines. Repeat the experiment, mixing a few drops of milk into the water, and see what happens to the intensity of the light. (Compare what you see with Experiment 14, "Why Is the Sky Blue?") How about repeating the experiment with a second hole, just a bit above or beside the first, with another piece of the drinking straw? Weld the two streams of water together (see *The World of Atoms and Our World: Cold, Heat, and Giant Bubbles*, Experiment 14, "Tying a Knot in a Stream of Water") and follow the path of the light. Try also using plastic tubing filled with water to guide light. Hint: Attach pieces of clear plastic to the tubing ends.

FUN FACTS

When light propagating in a transparent medium with a higher refractive index, like water, approaches the interface with a medium of lower refractive index (like air), it is partially reflected and partially transmitted. However, as the approach (incidence) angle becomes too shallow, there is no transmission at all, only internal total reflection (see also part B of Experiment 15, "Exploring the Laser Ray"). That is the way the optical fibers works. In practice, they are made of quartz or plastic and are used in optical communications to transmit light that carries information, using light signals to encode sound, images, and other kinds of retrievable data.

Reproduce the strobe light of discos, using your TV screen or computer monitor.

SUPPLIES

- computer monitor or TV screen
- variable-speed electric fan

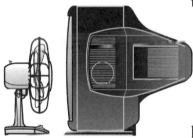

STEP BY STEP

Wiggle your finger back and forth really fast. Can you follow the movement of your finger? Turn on the computer or TV and repeat the same movement with your hand in front of the screen. What can you see? Now use a variable-speed electric fan between you and the screen and see what happens to the movement of the blades as you change the speed at which they rotate. Have you seen something similar in a disco with a strobe light? What can you say about the light from the screen? What happens if you speak loudly in front of the rotating blades of the fan?

FUN FACTS

The image formed on the screen of most modern (analog) computer monitors and televisions is produced by beams of electrons, which are tiny charged particles that sweep across the screen all the time, from left to right and top to bottom. Each electron that hits the phosphorus-coated screen creates a "spark" of light, and we perceive all the sparks together as a complete image. Since each point of the screen emits light only when an electron hits it, it flashes with a certain frequency. The same holds for the whole screen.

Playing with Light: Optics

Consider now a rotating disk with a mark at its rim (the electric fan blades) illuminated by a flashing light. If the frequency of the flashes is equal to the disk's rotation rate, the mark on the disk is illuminated always at the same spot and hence appears motionless. Now, if the flashing frequency is *less* than the disk's rotation rate, the light flashes again *after* the mark has completed a revolution. In this case, the mark appears to turn "forwards" (in the same direction it is actually turning). Similarly, if the flashing frequency is *greater* than the disk's rotation rate, then the light flashes *before* the mark can make a complete revolution. This repeated action makes the mark appear to turn backwards. (This always happens in old Western movies with the wheels of stage coaches.) In either case, the mark seems to travel a shorter distance between two consecutive flashes than it actually does, giving the impression that it is slowed down. This optical effect is called the *stroboscopic* effect (stroboscope means "whirling watcher" in Greek). Don't be disappointed if you can't see the strobe effect with a liquid crystal screen. Look at the screen from the side or use a polarizer held near your eye. Turn the polarizer and see what happens. You can start with the liquid crystal face of a digital watch.

FRACTAL CHRISTMAS

Find out more about the meaning of Christmas.

SUPPLIES

- 4 identical mirrored-glass balls (Christmas tree decorations)
- flashlight
- masking tape

STEP BY STEP

Place the four balls in a pyramid, as shown. Use the masking tape to stick them to each other. Shine the light on them, as the illustration shows. Now you can share these unique images with your friends.

A STEP FURTHER

Replace the ball on top with one of a different size. Also, you can use different colored balls and see what happens.

FUN FACTS

You probably have already seen in garages, street crossings, and in shops, mirrors that produce images like the mirrored-glass balls. Because they shrink the objects, you can have a bird's eye view of the objects in front of them. When you have several such mirrors close to each other, the image of one ball is a smaller ball, which in its turn, produces another image still smaller, and so on. So, you end up with what is called a *fractal*.

Playing with Light: Optics

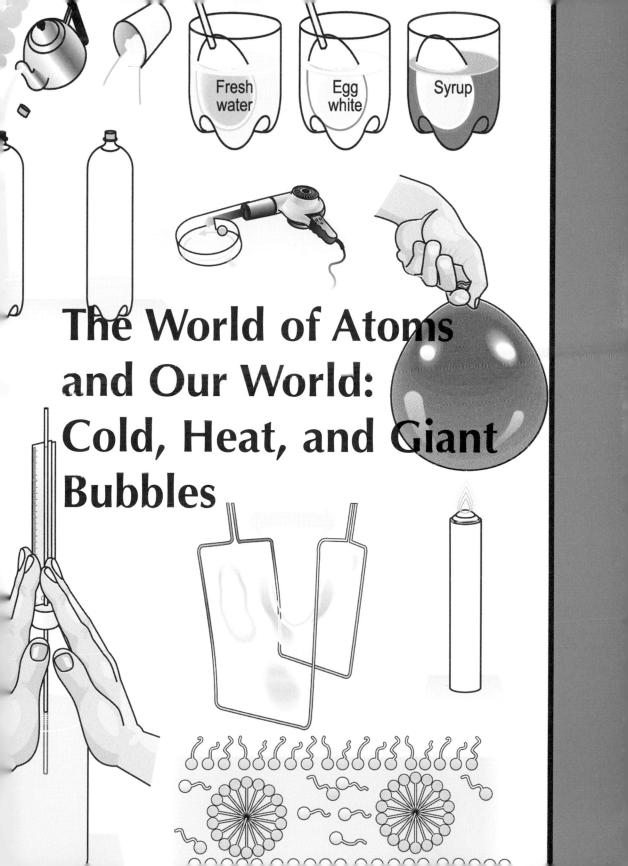

The World of Atoms and Our World: Cold, Heat, and Giant Bubbles

Get an idea of how atoms behave in gases, for example, in the air that we breathe.

SUPPLIES

- approx. 50 little Styrofoam balls about 1 in (2.5 cm) in diameter
- laundry or wastebasket with holes in the sides (no more than 1/2 in to 3/4 in (1.5–2 cm) wide holes
- cardboard
- hair dryer
- string

STEP BY STEP

Cut a circle out of cardboard, slightly smaller—about 3/4 in (2 cm)—than the basket, for a lid. Make a string handle and push it through two holes in the cardboard. Place all the Styrofoam balls inside the basket and close the lid. Holding up the basket on its side, make the *cold* air coming out of the hair dryer hit the basket, as the picture shows. Watch what happens to the balls. (Place the basket closer to the hair dryer if necessary.)

HINT

With the lid kept firmly in place inside the basket, increase the speed of the air coming out of the hair dryer (the "temperature" is increased). How do the Styrofoam balls behave with these changes?

The World of Atoms and Our World

Now, push the lid into the basket, leaving less space for the balls. How do they react?

FUN FACTS

A gas can be seen as a huge number of tiny balls moving in zigzags all the time—they knock into each other and the walls of their container. *Pressure* is then the net result of the banging of billions of billions of energetic tiny balls against their container's walls. If the space available for the balls decreases, they become crowded and knock each other and the container's walls more frequently, thus increasing pressure. *Temperature* is a measure of how energetic these tiny balls are. The higher the temperature, the more energetic they are. (This means that *on the average* the speed of the tiny balls will increase with heat—actually some will be much faster, others will be more sluggish, but most will be neither at the top nor at the bottom of the speed scale.) If temperature decreases, the tiny balls slow down and the bangs against the walls of the container become fainter and less frequent—so pressure goes down. Hence at a higher temperature, when the gas becomes more energetic, if its volume is kept constant, the frequency of overall collisions goes up and pressure increases.

CRUSHING CANS AND PLASTIC BOTTLES

It's possible to crush a can with a bucket of cold water!

SUPPLIES

- plastic bottle with screw-on top
- aluminum beverage can
- container with boiling water (like a teapot)
- container with cold water
- two long wooden sticks (for example, wooden spoon handles)

STEP BY STEP

Pour boiling water into the plastic bottle. Wait one or two minutes, then screw on the cap tightly. Now, dowse the bottle with the cold water and see what happens.

Pour enough water into the can to cover the bottom. Heat the can directly in a low flame or place it in a pot of boiling water, holding the can with the two wooden sticks. When the water inside the can begins to boil, turn the can quickly (and carefully!) upside down and let the top half submerge into the cold water. What happens to the can? Why?

SPECIAL WARNING

Keep the can extended far away from your body to avoid accidents.

HINT

What happens to the steam inside the plastic bottle or can when it is suddenly cooled off?

FUN FACTS

The temperature of the boiling water inside the bottle or the can affects the air inside, which becomes more energetic than the air outside.

As a result, the pressure inside the bottle or the can increases and the air flows out (until the pressure equalizes), decreasing the number of tiny jiggling balls left behind. When you screw on the cap and dowse the bottle with the cold water or suddenly let the top of the can submerge in cold water, the temperature immediately falls and the internal pressure falls, too, producing a partial vacuum. Now the tiny balls of the air outside do their job, since the outside pressure is higher than the inside pressure, and so both the bottle and the can are crushed.

---

3 ★ BENDING LASER BEAMS WITH HOT AIR

You can bend a laser beam by heating up part of its path.

SUPPLIES

- hair dryer
- laser-pointer
- laser-pointer holder (see part E of Experiment 15, "Exploring the Laser Ray")
- CD (with CD holder) or small mirror
- adhesive tape
- candle

STEP BY STEP

Switch on your laser pointer, fixed in the holder so that the laser beam hits a wall ("white screen"). Fix the CD or the small mirror to the wall with adhesive tape so that the laser beam is now reflected and projected to another wall (for example, opposite the mirror's wall) or to the ceiling. With the hair dryer switched on (hot air) very close to the light path, shake it to periodically intercept the laser beam with the flow of hot air. You can also use a candle instead of the hair dryer. In this case, the laser beam crosses the candle's flame, as shown. You can make the flame oscillate by blowing sideways. Look at the wall or ceiling where the laser beam is projected. How does it react to the shaking of the hair dryer or to the dancing flame?

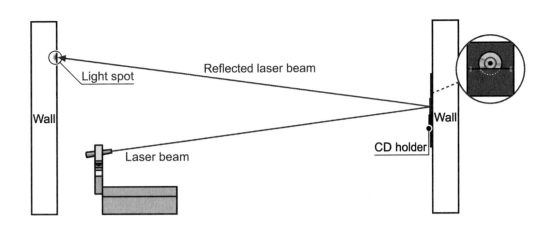

Reflected laser beam

Light spot

Wall

Laser beam

Wall

CD holder

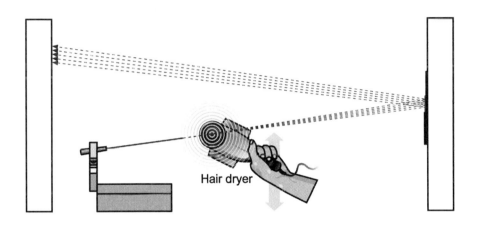

Hair dryer

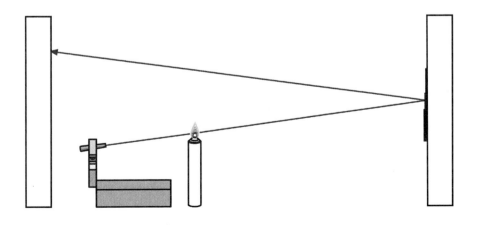

HINT ————————————————————

The longer the path of the reflected laser
beam, the greater the effect produced by the
hair dryer or the dancing flame. Can you
enhance the effect by using an extra CD or
mirror to further reflect the laser beam?

————————————————————

FUN FACTS

Hot air has a different refraction index than
"normal" air. This is related with the fact that
hot air is less dense than cold air. (The same
happens with water; see Experiment 11, "Car
in the Sun: Greenhouse Effect and Solar
Heater.") When you open the door of a hot
oven, you feel a stream of hot air coming up.
This means that when you intercept the laser
beam with hot air, you change its path. Have
you also noticed the trembling images you see
over the roof of a parked car on hot summer
days? Mirages in the desert or over hot asphalt
also arise from a varying index of refraction
due to differences in the air temperature. (At
very low angles, the light rays appear to come
from beneath the road, as if they were reflected in a puddle.) You may have noticed that the
stars seem to twinkle at night. The temperature of the air varies in time due to air currents
in the atmosphere. As a result, the light rays
from point light sources, like stars, undergo
random deviations (refractions) as they pass
through the atmosphere. On the Moon, for
example, you would not see stars twinkle
because there is no atmosphere there. The light
coming from distant places (buildings and
public illumination) also display similar effects. Besides variations of the air temperature,
pollution (for example, dust particles suspended in the air) can also deflect rays coming from
distant light sources.

Putting jiggling atoms to work.

SUPPLIES

- soda can (full)
- wire
- metal container (like an empty sardine can)
- carton with aluminum lining
- stapler and staples
- white glue or superglue
- chalk
- rubbing alcohol
- sink or large bowl
- needle-nose pliers or wooden sticks
- water and a piston coupled to plastic tubing (see Experiment 24, "Hydraulic Elevator," under *Fun with Mechanics*)

CAUTION

This experiment requires the help of an adult.

STEP BY STEP

Carefully make a tiny hole in the full can (the steam engine), in the place indicated (see inset illustration), using a safety pin or small nail, and hold the can upside-down over the sink or a large bowl. The soda will spray out of the hole. Turn the can upright when the contents reach about one-third of the original volume. Cut two 3 1/2 × 3 1/2 in (9 × 9 cm) squares from the carton with aluminum lining, and glue them back to back so that both sides have the aluminum face. Use this square to make the windmill. If necessary, use staples to make the windmill firmer (see inset). Bend the wire into the shape shown to hold up the soda can and to serve as the axis of the windmill, which should come out from the side of the can. Place the chalk in the empty metal can (the furnace), keeping the chalk completely below the top of the can, and pour the alcohol onto the chalk until it can no longer absorb the liquid (until it is saturated). Place the steam engine (can) into its holder and carefully set fire to the chalk/alcohol mix. After a few minutes, when the liquid in the steam engine (can) boils and the steam escapes, you will see the windmill spin.

Steam Machine *

Hole

Steam engine
(soda can)

Steam engine
(soda can)

Liquid

Fan

Chalk with alcohol

Metal
container

Infographic: Cláudio Roberto

Staples

Chalk is a porous material. The alcohol poured over the chalk is absorbed into it to be released later, little by little as it burns. (You can make campfires the same way—charcoal is a porous material, too.) Use only as much alcohol as instructed above, **keeping the bottle far from the fire and far out of reach of children**. Use a pair of needle-nose pliers or two wooden sticks (like wooden spoon handles) to take the can off the fire when it looks as if it is almost empty. The can will deform if it touches cold water (see Experiment 2, "Crushing Cans and Plastic Bottles," in this part of the book). To put out the fire in the "furnace," **don't put water on it!** Instead, cover the sardine can with a piece of wood or cardboard. (Since the chalk is completely below the top of the can, this will smother the fire faster than the fire will burn the board. Using damp/wet cardboard might be somewhat safer, as long as it isn't falling apart and doesn't leave openings for air to get in.) Let it cool down completely before re-using it. To refill the can, squirt water with the piston into the tiny hole, until it's one-third full again. With use, the can will deteriorate, so at some point it will be necessary to use a new can.

CAUTION ─────────────

Never heat the steam engine with soda or water directly in a strong flame (like a gas stove flame or blowtorch) or other strong heat. The pressure inside the can could increase quickly and cause an explosion. Following the instructions above eliminates such a risk of accident.

FUN FACTS

In water, the molecules are loosely tied to each other so that they still have mobility. (Water flows!) When you heat up water, the molecules become energetic enough to break the bonds and move away, so water becomes vapor. The pressure produced by the vapor inside the can is higher than the pressure exerted by the air outside the can, whose molecules are less energetic. When the inside pressure becomes great enough, a hot vapor jet comes out of the small hole and hits the windmill, making it rotate. When most of the water is vaporized, the pressure inside the can decreases (and it's time to refill it). In this process, the energy associated with the random motion of the water molecules (caused by heat) inside the can is only partially transformed into kinetic energy associated with the ordered motion of the spinning windmill. It is possible, though, to completely transform "ordered" energy into disordered energy (heat). This occurs, for example, in electrical irons and toasters, when electricity is converted into heat. After all, nature seems to prefer chaos to order. How could you make your steam machine more efficient? What about generating electricity out of it? (Hint: Use an electric motor in reverse, coupled to the windmill, see Experiment 9, "Electric Motor," under *Electrifying Experiments: Electricity and Magnetism.*)

Build different models of steamboats and see how they compare to each other.

SUPPLIES

- 8 to 10 in (20–25 cm) of copper tubing, the kind that bends easily (available at most hardware stores and places that sell refrigeration equipment)
- flat piece of Styrofoam
- thin, flat paneling, about 1/16 in (2 mm) thick, of wood or other hard material that can be used for the boat
- wood glue or superglue
- PVC pipe, battery, or piece of broom handle
- candle 2 in (5 cm) long
- tub or large basin of water

STEP BY STEP

A. Angled Hull

Cut a piece of the paneling in the shape of your steamboat, as shown in the illustration. Using this as the pattern, cut a piece of flat Styrofoam to the same dimensions. Glue them together. Make two holes, as shown, in the steamboat's deck so that the copper tubing can fit in tightly. Use the piece of broom handle or a similar item to roll up the copper tubing into a tight spiral (four or five times around), keeping ends of 3 to 4 in (7.5–10 cm). Push the ends of the copper tubing through the holes in the deck with the spiral high enough to fit

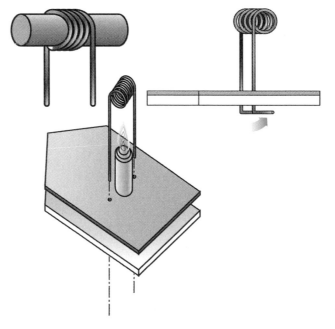

the candle underneath. Beneath the steamboat's hull, bend the copper tubing, as shown, so that the ends extend for a bit, parallel to the hull and in the same direction. Set the boat on the surface of the water (with the coil up), and light the candle. Wait a few seconds, and watch your boat take off across the water.

B. Circular Hull

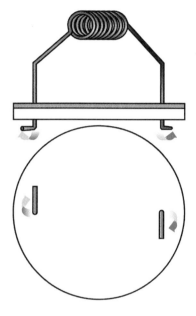

Make a steamboat with a circular hull, 2 in (5 cm) radius, and repeat the previous steps, with one exception. The ends of the copper tubing should be pointed in opposite directions, as shown in the illustration. Light the candle and see what happens. What relationship can you find between this experiment and Experiment 40, "Water Fountain," under *Fun with Mechanics*?

A STEP FURTHER

How about joining models A and B in a single prototype, using two copper tubing spirals and two candles? How do you think the new steamboat will move? You could also make a model A boat with additional candles (more power). Is there any advantage in using a second spiral?

FUN FACTS

Initially, there is some air inside the tubing, which is heated up by the flame of the candle. (Copper, like most metals, is a good conductor of heat. That's why the handles of large metal spoons used in the kitchen to stir hot liquids are made either of wood or plastic, which are poor conductors of heat.) The air pressure eventually becomes great enough to expel vapor and water in the tubing. As a reaction, the steamboat moves on. Since a partial vacuum is formed inside the tubing, water is sucked in. The flame heats up the water close to the flame, making it vaporize. The vapor pressure becomes high enough to expel the vapor and water in the tubing, and the whole process starts again. The steamboat is thus a steam machine that transforms heat into "ordered" motion.

ANOTHER STEP FURTHER: LASER MONITORING OF VAPOR OUTPUT

STEP BY STEP

Put some detergent in the water and stir it. The detergent softens the water, making it easier to form vapor bubbles. Also, the detergent molecules scatter light just as milk molecules do when mixed with water. Set the boat (model A or B) on the surface of the water and use the piece of wood to keep the boat fixed in place. Align the laser beam with the output of one of the two ends of the copper tubing, parallel to the hull. Light the candle and look from the side to see the vapor being expelled. What happens to the vapor output when you light an extra candle? How is the water sucked in? Does it affect the boat's performance?

FUN FACTS

As vapor is expelled, the refractive index along the laser beam changes, producing a

---

ADDITIONAL SUPPLIES

- clear glass baking dish almost completely filled with water
- laser pointer with holder (see Lens Made of a Drop of Water, part F of Experiment 15, "Exploring the Laser Ray," under *Playing with Light: Optics*)
- dishwashing detergent
- piece of wood with length greater than the width of the baking dish

contrast. (The laser beam is more scattered sideways, so you can have a snapshot of the expelled vapor.) Laser beams are used to measure the speed of fluids (for example, air, water, and blood), to monitor pollution in the atmosphere, and to control many industrial processes in hazardous conditions, like high-temperature environments. All this versatility makes the laser a very powerful tool to discover new things and to tackle a great number of practical problems.

6 BURN BALLOONS WITHOUT POPPING THEM

Air-filled balloons burst when you put them too near to the flame of a candle. But what if they are filled with water?

SUPPLIES

- party balloon
- candle
- matches

STEP BY STEP

Fill the balloon with water, being careful not to overfill it. Hold it by the neck and carefully suspend it over the candle flame.

- Does it burst because of the flame's heat? Filled with air, it would definitely pop.
- Why is water used in cooling systems (car engines and some other machines that heat up)?

Instead of a balloon, you could use a plastic bag or disposable cup.

- Why do we sweat in hot environments or when we exercise?

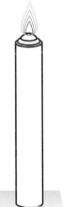

A STEP FURTHER

Fill a plastic cup halfway. Cover the cup with a piece of paper. Press the paper with your palm firmly to the edge of the cup and turn the cup quickly upside-down, securing it with your other hand. Pull your top hand away from the cup (the one holding the paper, not the cup). What can you see in common between this and Experiment 8, "Full Balloon with End Open"? Place the paper near the candle flame and see what happens.

FUN FACTS

In the absence of water, the heat provided by the flame goes mostly to the balloon, since air is a poor conductor of heat. As the balloon is heated up, bonds between the gigantic molecules ("macromolecules") that make it up are broken. Because the air pressure inside the balloon is greater than the outside pressure, as soon as a small hole is produced in the balloon, the air inside very quickly makes its way out, tearing the balloon further. When you fill the balloon with water, things change a lot. Water has a very special property. As it "sucks" heat, its temperature increases very little compared to other substances. So, water is an excellent cooling liquid. When you fill the balloon with water, the heat goes mostly to the water. Since hot water is less dense than cold water, as it is heated up, it moves upward and is replaced by cold water, which cools down the balloon at the point where you are heating it. Can you see now why we sweat when our bodies become hotter?

7

AIR AND WATER THERMOMETERS

Build a thermometer out of water and find out how it works.

SUPPLIES

- 1-pint (500-ml) glass bottle with top (you can also use the screw-on top from a plastic bottle)
- 1.15 ft (35 cm) of clear plastic tubing (for example, aquarium type) or use a transparent drinking straw
- 8 in (20 cm) ruler with measurements
- water
- poster paint or food coloring (for contrast)
- superglue

STEP BY STEP

Make a hole in the lid, slightly smaller than the tubing or drinking straw, for a tight fit. Run 4 in (10 cm) of the tubing through the hole. Pour water mixed with poster paint or food coloring into the bottle, filling it about two-thirds full. Screw the lid on tightly, with the end of the tubing down in the water. Use superglue to seal the tubing in the lid and to glue it to the ruler, in the position shown. Blow across the top of the tubing until a column of water rises about 1 in (2.5 cm). Press both of your hands to the top of the bottle and watch the column of water to find out how your thermometer works. How about repeating this experiment with the bottle totally filled? Now you have a new thermometer. Which of the two is more sensitive, the air or water thermometer?

A STEP FURTHER

Repeat the experiment with the air thermometer, holding the upper part of the bottle with the palm of one hand. With the other hand, hold a finger over the straw. What happens to the column of water in the straw? After a few seconds, take your finger off and see what happens. What relationship can you see between this experiment and Experiment

32, "Unwanted Ball," (under *Fun with Mechanics*) and Experiment 24, "Whirlpools," in this part of the book?

FUN FACTS

Our hands are warmer than the bottle and the air inside. (The average temperature of the human body is around 36° to 37°C.) As the top of the bottle is pressed with the hands, the air inside is heated up, becoming more energetic. As a result, the air pressure inside the bottle increases, forcing water to move into and up through the drinking straw. Eventually, the pressure exerted by the column of water formed in the straw balances the air pressure inside the bottle. Now, all you need to do is calibrate your air thermometer!

Water will increase its volume when warmed above 4°C (its temperature of maximum density), but not by very much. If we add to this that when the bottle is totally filled with water, the heat transferred to it by your hands will change its temperature very little (see previous experiment), it is not surprising that nothing noticeable happens.

8

FULL BALLOON WITH END OPEN

You can keep a balloon filled with air even with its end wide open.

SUPPLIES

- party balloon
- plastic bottle (preferably transparent)

STEP BY STEP

Make a hole in the bottle, at the bottom or on the side close to the bottom. Stretch the lips of the balloon over the bottle's opening and push the rest of the balloon inside. Then blow into the bottle to fill up the balloon. While you are blowing, air will come out of the hole you made in the bottle (put your finger there to check). When you finish blowing, cover the hole with your finger. Be careful not to make abrupt movements, especially if you are showing the experiment off to a large audience. Is the balloon with an open end still full? Uncover the hole and see what happens.

HINTS

1. With the balloon empty, suck the air out of the bottle through the little hole you made. In the final analysis, why does the balloon stay filled?

2. With the balloon filled and end open, place your palm over the top of the bottle to cover the balloon's open end, and uncover the hole you made. What happens to the balloon?

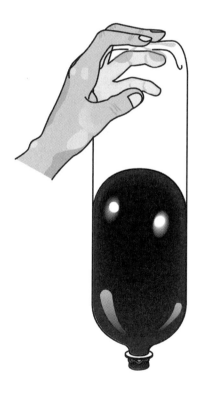

FUN FACTS

When you force the air out of a container, a partial vacuum is created inside. This happens because fewer jiggling atoms knock against each other and the walls of their container, and thus less pressure is produced. As you blow into the balloon, it expands due to its elasticity, forcing the air inside the bottle out through the small hole. When you stop blowing, the pressure inside the bottle becomes less than the atmospheric pressure. This pressure imbalance produces the force that keeps the balloon stretched when you cover the hole in the bottle with a finger and take your mouth away from the balloon. When you suck the air out of the bottle, you also create a partial vacuum inside the bottle, producing the same effect as blowing into the balloon, then stopping. This is basically the principle of vacuum pumps. Now, if you cover the top of the bottle and uncover the hole, the air pressure inside the bottle increases, pushing in the balloon's wall. Can you see how you end up with an ordinary filled balloon?

9 INVISIBLE HAND

Show that there are gloves even for invisible hands!

SUPPLIES

- rubber glove
- drinking straw
- large clear glass jar
- water

STEP BY STEP

Place the glove inside the glass jar and stretch it until it fits over the edge of the jar, leaving only a tiny space open for the drinking straw to fit in (see illustration). Fill the glove with water and, while the glove is filling up, place your finger near the open end of the straw and feel what happens as the glove fills. When the glove is full, take out the straw and stretch that part, too, over the edge of the jar, sealing it tightly. Pour the water out and see what happens. What relationship can you see between this experiment and the previous one?

HINT

Place your hand inside the glove and try to pull it out of the jar. What happens? Compare this situation with Experiment 1, "Jiggling Atoms," in this part of the book. What happens when you increase the volume by pulling the lid further out?

FUN FACTS

This is a variant of the previous experiment. In this case, however, you can feel the effect of a partial vacuum directly as you try to pull the glove out of the jar. This relates to space suit construction for astronauts. The first gloves made for astronauts, for example, expanded just like your "invisible hand." This made it harder for them to do work that required manual dexterity, and in general the suits were more tiring to wear while engaging in activity. You can imagine why if you did this experiment carefully. Great advances are still being made, but the search for a more comfortable space suit continues.

Find out how pneumatic tire valves work.

SUPPLIES

- party balloon
- Styrofoam ball or marble
- basin or large bowl of water
- hair dryer

STEP BY STEP

Place the ball inside the balloon (stretch the neck out until you can get the ball to pass through it). Blow it up, then hold the neck closed with your fingers while you turn the balloon until the ball gets close to the opening. Let go. What happens? What function does the ball serve?

Now place the filled balloon in a basin of water. Does it sink? Let the air out of the balloon, keeping the ball inside, and place it in the basin of water. What happens? Compare this experiment with Experiment 22, "The Submarine," under *Fun with Mechanics*.

A STEP FURTHER

Blow up the balloon with the Styrofoam ball inside. See what happens to the ball while you fill up the balloon.

Make an almost circular track out of a piece of poster board 8 × 24 in (20 cm × 60 cm) and point the hair dryer (cold air) at the opening of the circle (see the illustration). Place a Styrofoam ball in the middle of the track and see what happens. Put more balls inside. Can they get out? Change the shape of the track

Marble

and repeat the experiment. Also, think about how you can change the direction of the airflow from the dryer. What relationship can you see between this experiment and the ball inside the balloon?

ALTERNATIVE

Try placing the ball inside a large mug with the stream of (cold) air pointed toward the inner wall. Then try out various directions for the stream of (cold) air.

FUN FACTS: TIRE VALVES

Since the pressure inside the balloon is greater than outside, the air inside forces the Styrofoam ball or marble out. However, the neck of the balloon is less elastic (see Experiment 5, "Pierce Balloons without Popping Them," under *Fun with Mechanics*), and the ball gets stuck to it, preventing the air inside the balloon from coming out. In a real tire valve, there is a small case with a small metallic ball confined inside. When the external pressure is greater, the ball is pulled, letting the air in. As the pressure gauge is switched off, the pressure inside the tire becomes higher than the outside pressure, and the ball returns to its original position, preventing the air inside the tire from escaping. The balloon corresponds to the tire, its neck to the confining case, and the Styrofoam ball or marble to the metallic ball.

The small ball is forced against the inner walls of the balloon as you blow air inside (the increasing pressure inside makes the balloon expand). However, the balloon is not a perfect ball, so the air flow produces a whirlpool inside, as the movement of the Styrofoam ball demonstrates.

11 | CAR IN THE SUN: GREENHOUSE EFFECT AND SOLAR HEATER

This is your big chance to build a solar heater and find out how it works.

SUPPLIES

- cardboard box with a depth of 1 1/2 to 2 3/4 in (4–7cm), no lid
- piece of glass big enough to serve as a lid to the box
- black matte construction paper or black poster paint
- lamp with standard filament light bulb, 100 to 150 watts
- thermometer (found at a pharmacy)

STEP BY STEP

Line the inside of the cardboard box with black construction paper, or paint it black and let it dry. Put the thermometer inside and the glass on top. Turn the lamp on and place

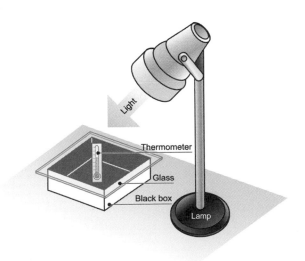

Light

Thermometer

Glass

Black box

Lamp

it close to the glass top. Watch how the temperature inside the box changes when the "Sun" (light bulb) is present.

FUN FACTS

The light passes through the glass and is absorbed by the dark sides of the box. The sides then emit infrared light, the kind that your eyes can't see but your skin feels, and it is strongly reflected by the glass. The inside of the box acts like the inside of a car left in the Sun with the windows closed. "The greenhouse effect" (slow heating up of our planet) is largely attributed to the emission of carbon dioxide (CO_2) by vehicles, factories, and fires. The CO_2, like the glass, reflects infrared rays, and the surface of the Earth acts like the inside walls of the box, absorbing sunlight and emitting infrared light.

SOLAR HEATER ★★

SUPPLIES

- 2-quart (2-liter) plastic bottle
- clear plastic tubing, around 10 ft (3 m) long, with an external diameter of 1/4 in (6.35 mm) and wall thickness of 1 mm, to form the coil
- 2 pieces of plastic tubing, with an internal diameter just a bit smaller than the external diameter of the other tubing (coil)
- matte black paint
- paintbrush
- standard 100-watt filament bulb with socket, cord, and plug
- poster board
- aluminum foil
- superglue
- empty vegetable can large enough to hold the funnel with bulb with space left over (see figure)
- masking tape

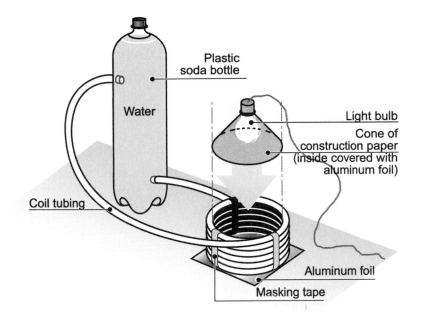

Plastic soda bottle

Water

Light bulb

Cone of construction paper (inside covered with aluminum foil)

Coil tubing

Aluminum foil

Masking tape

STEP BY STEP

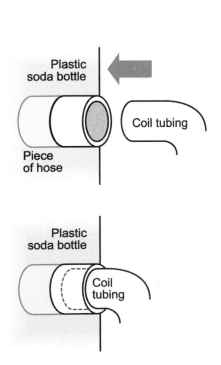

Plastic soda bottle

Coil tubing

Piece of hose

Plastic soda bottle

Coil tubing

1. Make a funnel-shape (cone) out of the poster board to go around the socket, making sure it will not touch the bulb (which will be very hot), and cover the inside of the cone with aluminum foil, as the picture shows. (The more reflective side should be exposed to better reflect the light.) Wrap the tubing around the can about seven times in a spiral, making a coil, and tape it into that shape with the masking tape. Take the can out of the center of the coil. Fill the tubing with water and close off the ends. Paint the inside of the coil with black paint (see the illustration). Place the cone on top of the coil, so that the bulb is right in the center of the coil but does not touch any side. You might want to light the bulb to dry the paint quicker (exterior car paint is set the same way in the factories). The water prevents the tubing from melting

(see Experiment 6, "Burn Balloons without Popping Them").

2. Make two holes in the plastic bottle, as shown in the blown-up illustration, putting the two small pieces of the slightly larger tubing into the holes, and seal them in the holes with the superglue. Fill the bottle with water until the level passes the lower hole, uncover the end of the tubing coming from the bottom of the coil, and fit that end into the lower hole. Continue to fill up the bottle until the top hole is covered by water, and place the other end of the tubing from the coil without its cover into the bottle. Then screw the lid of the bottle on tightly.

3. Place the coil on top of a square covered with aluminum foil with the more reflective side up. Use the cone to cover the top part of the coil. Turn on the light bulb, and after 10 or 15 minutes, touch the bottom and top of the bottle.

What can you conclude about the density of hot water versus the density of cold water? What's the importance of the matte black paint? Look at the next experiment, "Can Competition: Which Heats Up and Cools Down Faster?" So, how does the solar heater work?

SPECIAL WARNINGS

The light bulb should not touch the plastic tubing. If it does, the tubing could melt (despite being filled with water) and spray hot water. The hot glass of the bulb, when hit by the spray of water, could implode and send out glass fragments.

The black parts of the coil absorb light very efficiently and transform light energy into heat (the energy associated with disordered vibrations of atoms in the tubing). The temperature of the coil then becomes higher than the temperature of the water inside it. The water thus "sucks" the heat generated to adjust its temperature to that of the coil. As the water heats up, it expands and its density decreases. The heated water then moves up to the top of the bottle and is replaced by the denser cold water. You can check this by pressing your hands to the top and then to the bottom of the bottle. (It may take a while before you feel some difference in temperature.) This process is repeated until all the water is warmed up. Can you improve this solar heater? Are you ready to build an inexpensive solar heater efficient enough for domestic use?

12 CAN COMPETITION: WHICH HEATS UP AND COOLS DOWN FASTER?

Turn simple aluminum cans into Olympic champions.

SUPPLIES

- 2 aluminum beverage cans
- black paint
- paintbrush
- hot and cold water

STEP BY STEP

Paint the outside of one of the cans black and wait for the paint to dry. Fill both cans with hot water (preferably boiling). Wait 10 minutes and handle the two cans. Which is *less* hot? Now empty out the two cans and fill them up again with very cold water. After 10 minutes, handle the cans to find out which

one is *less* cold. You can also watch the temperature changes with two thermometers (found in pharmacies; SEE CAUTION BELOW). Now, can you say which the winner in each case was? What if you paint the cans red, yellow, green, and blue and put them to the test?

ALTERNATIVE

With a lens like the one you made of water (see Experiment 5, "Lenses Made of Air and Water," under *Playing with Light: Optics*), focus sunlight on a sheet of white paper and see what happens. Color a certain area of the paper with black ink and focus the light on that area. What is the difference between the white and black areas? (Be careful to avoid starting a fire!)

FUN FACTS

When we see a green object, like the leaf of a plant, that means that all colors are absorbed by the object except green, which is reflected. In the case of a black object (a "blackbody"), it efficiently absorbs most of the light hitting it. All the energy carried by the light goes into the object to increase the vibration of its atoms. In the case of metals, this absorbed energy also goes to "free" electrons wandering around inside the metal. These electrons are responsible for the electrical conduction in a wire, for example, when you connect its ends to a battery. In an incandescent lamp, the tungsten filament, which is black, emits light when it is heated by an electrical current. This holds for all blackbodies when they are heated, as you can demonstrate with the cans. (Most often, the light emitted by a blackbody is infrared light, which is not visible. This is also true for the filament lamp. That is why it is not energy efficient for illumination.) If a body is efficient for taking, it should be equally efficient for giving! It's only fair. Can you now understand what happens when you do the experiment with cold water?

13 FOG-PROOF MIRRORS

When you take a hot shower, your mirror always gets foggy. Let's get this problem sorted out.

SUPPLIES

- standard 100-watt filament bulb with socket, cord, and plug
- plastic container (without lid), big enough for the bulb
- aluminum foil
- flat mirror with the edges sanded smooth, big enough to serve as a lid to the box
- matte black paint
- paintbrush

STEP BY STEP

Cut a hole in the corner of the plastic container to tightly insert the bulb socket. Cover the interior of the container with aluminum foil.

> **CAUTION**
>
> Make sure that the lamp's electrical connections do not touch the foil, or the lamp will be short-circuited.

Paint the back of the mirror with matte black paint and wait until the paint is dry. Let your mirror get foggy and then switch the light on. It will shortly shine again! To avoid overheating, don't leave the bulb turned on for a long time, just a few minutes.

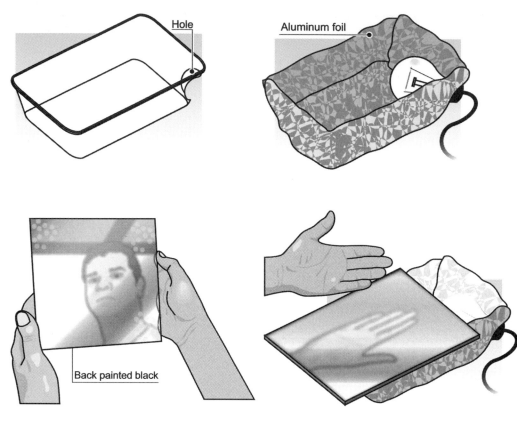

FUN FACTS

The light emitted by the bulb is mostly infrared light. (We don't see it, so most of the energy is lost for illumination purposes.) It is reflected by the aluminum foil and absorbed by the back of the mirror that is painted black (see Experiment 12, "Can Competition"), which gets hot enough to kick off the vapor stuck to the mirror. (As the temperature of the mirror increases, the water molecules attached to the mirror surface become energetic enough to break their bonds and move away.)

TYING A KNOT IN A STREAM OF WATER

Have you ever imagined sticking streams of water together? It's so easy!

SUPPLIES

- plastic bottle or waterproof cardboard box (like a juice carton)
- water

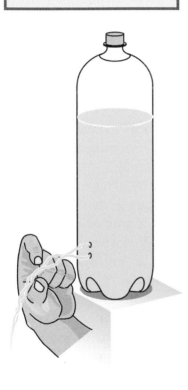

STEP BY STEP

Do this experiment inside a large basin or bowl, or in a sink, so as not to get water all over the floor. Make two holes 3/8 to 3/4 in (1 to 2 cm) apart in the container, as shown. Over the sink or basin, fill the container with water. (To keep the water from escaping, hold your fingers over the holes.) Next, take your fingers off the holes. Now, bring the two streams of water together with your fingers, and see what happens. Why does water stick to water?

FUN FACTS

On the surface of water, the molecules are tied loosely to their neighbors along all directions, except above, where there is only air. So they feel a net force pulling them down. Since the space below is already occupied, an inward tension is produced. This is called *surface tension.* The larger the surface, the greater the energy stored as tension. A smaller surface means less tension. When you bring the two streams together, which you can think of in terms of two cylinders, you produce a single stream with less surface area than the sum of each individual stream. The water, as other fluids do, prefers to be as relaxed as possible. When your hair is wet, the strands stay very close together to make the area in between as small as possible. Keep cool, the water in your hair is most relaxed!

SOAP SADDLES? YOU ARE JOKING!

They might be useless for horseback riding, but they will definitely catch your fancy.

SUPPLIES

- 2 pieces of solid wire, about 2.5 ft (77 cm) long
- 2 pieces of plastic tubing (aquarium type), 5 1/8 in (13 cm) long
- soap solution: approximately 2 quarts (2 liters) water, 250 ml dishwashing detergent, and 100 ml glycerin (found in pharmacies)—try other proportions, too
- shallow container large enough to hold the soap solution

STEP BY STEP

The first step is to make the soap solution. Some detergents contain glycerin, so you may not need to add glycerin to your solution—check the ingredients on the detergent's label. There is no universal recipe for soap solutions, since they depend on the local water and on the kind of detergent you use. So, invent your own recipe for this experiment and mix your solution in the shallow container. Each component has a function: the detergent is to make water more "elastic," and the glycerin is to prevent it from evaporating too quickly.

Fold each piece of solid wire as indicated in the figure, saving about 6 in (15 cm) for the handle. You will get a pair of two-legged chairs, where the handle is the chair back. Make a hole just large enough to pass through two ends of solid wire 3/8 in (1 cm) from one end of each plastic tubing. Drill two smaller holes all the way through from one side of each plastic tubing to the other, 3/8 in (1 cm) from the other end of the tubing, enough to pass just one end of solid wire. Now drill two similar holes 3/4 in (2 cm) apart from the previous ones. Insert each pair of "chair backs" into the larger hole in the corresponding tubing. Pass each wire end through one of the smaller holes, fold the wire, and then insert each end back into the tubing through one of the smaller holes. In

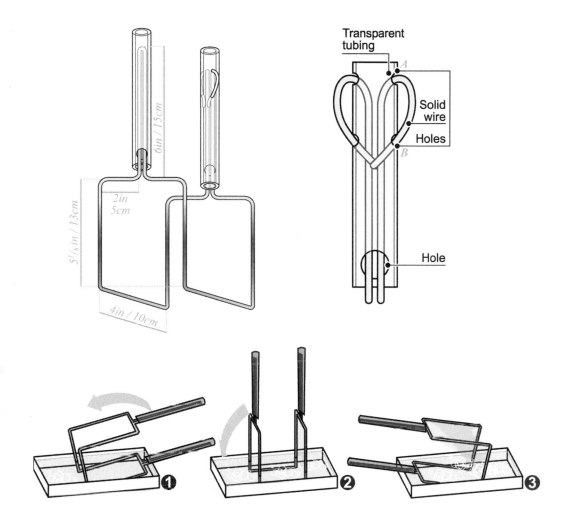

this way. the handles are safe. (All wire ends are inside the pieces of plastic tubing, which also keeps the two halves of the saddle support tightly together.) Hold each handle with one hand and plunge it into the soap solution. A colorful saddle, almost magically, will form as soon as you take the support out. Now adjust the saddle to your liking by pulling apart or pushing together the handles, and enjoy yourself.

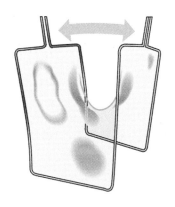

The surfaces of the soap saddles anchored in the wire frame of your saddle support have the smallest possible areas for each position of the handles. The soap tries to stretch as little as possible. Other surfaces would mean more tension for the soap, which is basically an elastic skin, just like party balloons. If you poke a balloon with a pin at a stretched point, one of the reasons it pops is this reduction of tension by making the surface smaller. The hole becomes bigger, allowing the stretched surface to become more relaxed! (See Experiment 5, "Pierce Balloons without Popping Them," under *Fun with Mechanics*, and Experiment 14, "Tying a Knot in a Stream of Water.") Try to poke the soap film of the saddle with a pin. (Try it first with the pin dry and then plunge the pin into the soap solution before using it to poke the film.) In which part of the saddle is the soap film most stretched? Do soap films display birefringence? What about adding handles to the horizontal wire pieces at the bottom of the wire frame? They would give you extra control over the form of your soap saddle. (In this case, you would need a container for the soap solution deep enough to accommodate the new handles.) Can you concoct other cool models of wire frames to anchor soap films?

RACQUETS AND TENNIS BALLS MADE OF SOAP

Get ready for a new kind of tennis.

SUPPLIES

- circle, 4 3/4 in (12 cm) in diameter, made of wire or plastic, with a handle (frame)
- tube from a ballpoint pen (without the ink cartridge) or a piece of drinking straw
- soap solution from previous experiment

STEP BY STEP

Submerge the wire frame in the solution. When pulling it out of the solution, you will see the film of soap fill up the whole circle. This is your racquet! Hold the pen tube by one end and dip the other into the solution for a few seconds. Pull the tube out and blow into the other end, forming a soap bubble. This is your tennis ball! Now you can practice your serve.

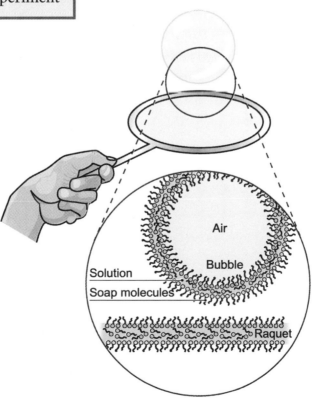

Air

Bubble

Solution

Soap molecules

Raquet

You can play tennis in this new way thanks to the properties of certain molecules that are found in detergent, called surfactants. These molecules can be drawn as little tadpoles, each with a "head" and a "tail." If the head of the molecule loves water (for this reason, it is called *hydrophilic*), the tail hates it (called *hydrophobic*). From the illustrations, you can see that in the soap ball, as much as in the racquet, the tails of these detergent molecules are on the outside of the film of soap. The term surfactant comes from the fact that when a detergent molecule is put in water, as many as possible will crowd to the surface so that the heads can stay in the water while the tails stick out into the air. Surfactants therefore mostly affect the surface of water. It is these molecules that make soap bubbles stable. Without them, the bubbles would spontaneously break apart into tiny water droplets. The ball and racquet don't stick to each other because the "tails" of one want to avoid touching the water that is on the film of the other! In reality, the surfactant molecules are much smaller than as illustrat-

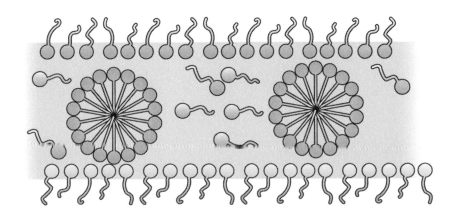

ed here, about 1,000 times smaller than the thickness of the bubble wall. Bubble walls are typically a few micrometers thick (the diameter of a human hair is around 50 to 100 microns, 1 micron = 10^{-6} m), while the molecules are just a few tens of nanometers long (1 nm = 10^{-9} m).

As you can see in the picture, there are also detergent molecules inside the water of the bubble wall, some of them by themselves and others in little balls known as *micelles*. In the micelles, many molecules ball up with their tails bunched up inside the ball and their heads on the ball surface. The tails thus avoid water, while the heads get to touch it. This is a handy way of storing soap inside the bubble wall. If the bubble is stretched, the molecules on the surface spread apart as the surface is enlarged. The additional molecules stored in micelles replenish the surface, so it remains nice and stable. When you use detergent with water to remove dirt from clothes and dishes, for example, the tails of the detergent molecules stick to the dirt, which tends to be greasy or oily, while their heads prefer the water. So, when you rinse the dish you are washing, the water takes the dirt and grease along with it, and the dish becomes clean.

How would you like to create ever-changing paintings on a film of soap?

SUPPLIES

- soap solution from Experiment 15, "Soap Saddles? You Are Joking!"
- rubber band
- flashlight

STEP BY STEP

Stretch the rubber band between the fingers of both hands (see picture), and dip it into the soap solution. In a dark place, ask someone to shine a flashlight on the soap film inside the rubber band (your painting). Gradually stretch the rubber band and watch the light reflected on the film. What relationship is there between the thickness of the film and what you can see on it?

FUN FACTS

When light hits the near surface of the soap film, it is partially reflected and partially transmitted through the soap film (see Experiment 9, "Magical Theater," under *Playing with Light: Optics*). The transmitted light, in its turn, is also partially reflected back at the far surface toward and (partially, again) out through the near surface. The two waves (the first reflection and the second coming back out of the film) superpose. Since each color has a different refraction index (different speeds of propagation inside the film), it takes a different amount of time for each component to cross through the film and return. If for a certain light component both waves look the same (are "in phase," as in figure A), they add up, producing a bright color. By contrast, when the electric field of the two waves cancel each other (the waves are "out of phase," as in figure B), the corresponding color vanishes. Since the film thickness varies over the film, in some places a certain color will stand out as it is reinforced, so you see areas with different colors.

As you stretch the rubber band, the film is also stretched and becomes thinner. This changes the color pattern resulting from the superposition of the two waves related to each color coming out of the near surface of the soap film. You can thus produce different cool paintings for your delight.

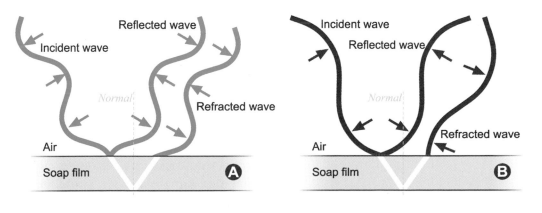

TWO-DIMENSIONAL VORTEX

Would you like to make vortexes on a film of soap?

SUPPLIES

- soap racquet (see Experiment 16, "Racquets and Tennis Balls Made of Soap")
- soap solution from Experiment 15, "Soap Saddles? You Are Joking!"
- drinking straw
- flashlight

STEP BY STEP

Use the flashlight to illuminate the soap film in the racket and position the straw near the film. Blow into the straw, as indicated in the picture, and watch the colorful vortexes that you make on the film.

A STEP FURTHER

Repeat the experiment using two or more straws at the same time. Can you make two vortices interact? Ask someone to produce a vortex rotating in the opposite direction of yours and check if they cancel each other out.

FUN FACTS

When you blow the soap film, it is dragged with the air flow, and a vortex is formed. Since the film thickness is not the same, the areas of different thickness move with the vortex. Considering that the film is very thin, you can say that the vortex is confined to a surface. You can swirl the colors because the areas of different thickness are dragged along with the air flow. (See Experiment 17, "Flexible Colors.") In this way, you can produce unique, dynamic paintings.

PASS THROUGH A SOAP FILM WITHOUT POPPING IT

Learn to conquer barriers.

SUPPLIES

- soap racquet (see Experiment 16, "Racquets and Tennis Balls Made of Soap")
- soap solution from Experiment 15, "Soap Saddles? You Are Joking!"
- marble or Styrofoam ball or billiard ball (try other objects)
- piece of thin wire or a paper clip

STEP BY STEP

With your finger dry, try to touch the soap racquet. What happens? Next, dip your finger in the soap solution for a few seconds, then try to poke a hole in the soap racquet. What happens then? Now, submerge the marble (or other ball) in the solution and let it fall onto the surface of the soap racquet. Does the film break? Try letting the ball fall from different heights, thus hitting the film with different velocities. Repeat the experiment using other objects.

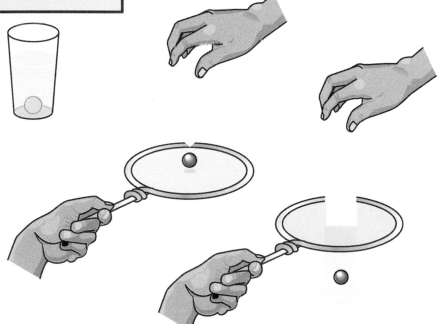

When you plunge the wire frame into the soap solution to form the soap racquet, the frame becomes wetted. This is a precondition for the soap film becoming anchored on the wire frame. The soap film thus becomes much thicker around the frame than in the center of the racquet, where its thickness is of the order of a few microns (one micron is around one hundredth of a hair's diameter). If you poke the soap film in the center of the racquet with a dry piece of thin wire or the end of a paper clip, there is not enough solution in there to wet the wire/clip, so the film will not be able to get anchored. Since the film is a stretched elastic membrane due to the surface tension, it tears apart as soon as you touch the film with something dry. (Now try to pierce the film with a dry thin wire or paper clip close to the wire frame.) The marble wetted ("dressed") with soap is recognized as part of the soap film as it passes through the film. (The soap film becomes anchored on the film covering the ball.) In the case of a "naked" ball, it produces a hole in the very thin soap film, because there is no way for the film to get anchored on the ball's surface.

You might wonder about the finding that in the atomic world an electron can cross a "barrier" without changing its characteristics, similar to what happens when the "dressed" ball passes through the soap film. This phenomenon is called *tunneling*. It is as if the electron passes through the barrier by carving out a tunnel, which closes up behind it.

NON-CUTTING SCISSORS

Try to cut a soap film with a pair of scissors.

SUPPLIES

- soap racquet (see Experiment 16, "Racquets and Tennis Balls Made of Soap")
- soap solution from Experiment 15, "Soap Saddles? You Are Joking!"
- scissors

STEP BY STEP

Dip the scissors in the solution, as indicated. Now try to cut open the soap film in the soap racquet.

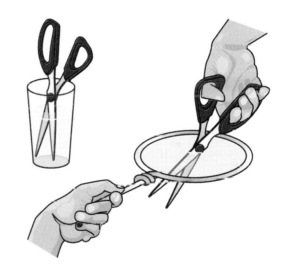

FUN FACTS

This experiment is, in fact, a variant of the previous one. The scissors wetted with soap solution act just like the ball, so the film becomes anchored to it. As you try to "cut" the film, no apparent damage to the film is produced, as before. Since the soap film is an elastic skin, it becomes more stretched as the scissor blades move apart and relaxes as they come together. (Can you use the wetted scissors to produce "flexible colors"? See Experiment 17, "Flexible Colors.") Now, try it with "naked" (dry) scissors. Will the trick still work? Why?

GIANT SOAP BUBBLES AND FILMS

Explore the fascinating world of giant bubbles.

SUPPLIES

- wooden pieces 3/8 to 5/8 in (1–1.5 cm) thick, in the following dimensions: 1 piece 4 in x 2.3 ft (10 x 70 cm); 2 pieces 4 in x 2.5 ft (10 x 75 cm), and 1 piece 10 in × 2.3 ft (25 × 70 cm)
- PVC pipe, 4 in (100 mm) in diameter and 0.7 yd (64 cm) long
- 1/2 inch PVC pipe, 1.9 ft (58 cm) long (a broom handle might serve)
- 2 more pieces of 1/2 inch PVC pipe, 5/8 in (1.5 cm) long and cut into equal halves
- 2.6 yd (2.4 m) of nylon fishing wire, 0.5 mm thick
- screw-in hook
- 2 small wood screws
- corks or modeling clay
- superglue
- hard plastic, like a CD no one uses
- soap solution from Experiment 15, "Soap Saddles? You Are Joking!"
- flashlight

STEP BY STEP

Frame

Make the frame as the illustration shows, and screw the hook into the top, exactly in the center.

Trough

With a hacksaw (usually for metal), cut lengthwise down the center of the 4 in (10 cm) PVC pipe. Make a trough by gluing a piece of hard plastic on either end of the half-pipe. If you use a CD, cut it from the edge to the center with heavy-duty scissors (but not good sewing scissors). Take the four halves of the 1/2 inch thick PVC pipe, and place two on either side of the trough. Glue them in place as supports so that the trough can't roll over when it is filled (see the illustration below).

PVC Soap-Stretcher

Near each end (same distance from each end) of the 1/2 inch PVC pipe, drill a centered hole from one side to the other so that the nylon fishing line can pass straight through. This line will be the guide for the top and sides of the film. Beside each of these holes, make two more holes (again, the same distance from each end of the pipe). Cut a piece of line a bit longer than the length of the PVC to serve as

Giant Soap Bubbles and Films

Giant soap films and bubbles
Knot
Frame
PVC soap-stretcher
Cork
Holes
Fishing-line guide
Fishing-line guide
½ CD
Trough (PVC 4 in / 100 mm)
PVC support pieces
Film
½ CD
Fishing-line guide
Trough
Small screw
Solution
Soap solution

Infographic: Cláudio Roberto

a handle, and pass it through these holes, making a knot (larger than the hole) in each end. Stop up the ends of the PVC pipe with wine corks or modeling clay and, if using corks, cut the ends off with a craft knife, exactly at the end of the pipe. If PVC pipe is not available, you can use a broom handle.

Guides for PVC Soap-Stretcher

Cut two pieces of the nylon fishing line, 3.6 ft (110 cm) in length to use as guidelines. Screw the small screws into the bottom of the trough, near the ends (again, the same distance from each end), as shown. The positions of the screws (which will secure the guidelines at the bottom) should match the positions of the guideline holes in the soap-stretcher pipe (top) and the holes in the top support of the wooden frame. Tie the ends of the lines to the screws, and pull them up high through the guideline holes in the soap-stretcher pipe, and finally to the guideline holes in the top support (see the inset illustration). Stretch the lines tightly and knot them above the frame top.

Finally, the Giant Bubbles

Fill the trough with the soap solution, and completely submerge the soap-stretcher in the solution, laying it at the bottom of the trough. Wet the guidelines with the soap solution. Now, you only need to pull the soap stretcher carefully upward and hang its handle on the hook. Blow on the film and have fun with the bubbles. When the room is dark, you can shine a flashlight on the film, varying the angle of the light as it hits the film. Try to make waves on the film with the flashlight shining on it by shaking the PVC soap-stretcher. The polarizing filters from Experiment 13, "New Discoveries with Polaroids," under *Playing with Light: Optics*, offer yet more opportunities. The universe of bubbles is yours to explore to the fullest!

TWO STEPS FURTHER

A. Tunneling

Try to throw a small Styrofoam ball covered in the soap solution at the soap film, as in Experiment 19, "Pass Through a Soap Film without Popping It." Try launching the ball at various speeds and repeat the experiment with other objects.

B. Windows

Bend the handle of the soap racquet (see Experiment 16, "Racquets and Tennis Balls Made of Soap") so that it is perpendicular to the circular wire frame. Wet the circular frame with the soap solution. Touch the giant soap film gently with the wetted frame so that the film becomes anchored to it. Now, with a dry finger, poke a hole in the soap racquet. That's your window. Ask someone to produce waves on the film by shaking the PVC soap-stretcher. You can easily move the window sideways. How is the wave pattern affected by the window at different positions? You can also try other windows with different forms and sizes. By changing the position of the window, you are changing the "boundary conditions," as mathematicians and physicists call it. The advantage here is that you don't have to do any mathematics—

just move the window(s) around as you please and produce different cool (stationary) wave patterns.

FUN FACTS

The soap film is an elastic membrane. It is actually much more elastic than water due to the action of the soap molecules (which soften water). Since you can produce waves on the surface of water and on the elastic skin of drums, why can't you also do the same with soap films? Since the film is anchored on the lateral guidelines, the waves produced when you shake the soap-stretcher are reflected at the fixed film's boundaries. The incoming waves and the reflected waves combine to produce the beautiful patterns you see. (Other patterns are not possible because they correspond to destructive superposition of waves.) Since the thickness of the film is not uniform and changes as it is further stretched, its colors vary across the film (see Experiment 17, "Flexible Colors," in this part of the book). Boundary conditions define the wave pattern, since the reflection of waves occurs at the boundaries. You can also produce waves by shaking the soap film at its top perpendicularly to the film with a wetted piece of PVC pipe about 4 in (10 cm) shorter than the soap-stretcher and see how the window affects the propagation of waves consisting of line wave fronts regularly spaced.

22

SPEEDING UP WATER DROPLETS

Pave the way for a water droplet and make a champion out of it.

SUPPLIES

- 1 piece of glass with the edges sanded smooth (you can also use the outer bottom of a serving dish made of glass)
- drinking straw or dropper
- water repellent (used on car windshields)
- paper towel or flannel
- water

STEP BY STEP

Put the straw into the water and stop its top with your finger (to get some water). Take the straw out and unstop it briefly over the glass, held tilted down. A drop of water will fall and hit the glass. Does it slide down the glass fast enough? You can speed it up by further tilting the glass. Now, dry the glass or use the other side of it. Carefully pour a bit of water repellent on the paper towel and spread it all over the glass surface. If necessary, repeat this procedure. Wait for a while (read the instructions that come with the water repellent about the necessary waiting time) and then remove the excess of water repellent from the glass with a new piece of paper towel. Keeping the glass tilted with its coated surface up, drop a water droplet on it and see if the droplet performance improves. Is it now prepared to become a champion?

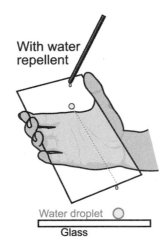

With water repellent

Water droplet

Glass

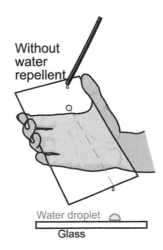

Without water repellent

Water droplet

Glass

A STEP FURTHER

Leave a strip of the glass uncoated (cover it with a narrow strip of paper before spreading the water repellent) and organize an obstacle race for smaller and bigger droplets. You can also use a comb to produce a "speed bump" road for water droplets. Feel free to explore the whole range of possibilities.

FUN FACTS

The water repellents used on car windshields are formulated with a transparent polymer that fills the microscopic pores of glass with hydrophobic molecules that force rain, sleet, and snow to bead up and roll off, thus improving all-weather visibility. If you closely examine the water drop on the coated glass placed horizontally, you will notice that the drop is almost a ball. This is the shape with the *minimum surface area* for a given volume. That means minimum surface tension (see Experiment 14, "Tying a Knot in a Stream of Water"). Since the drop now has very little contact with the glass surface (which means less friction), it can move much faster. That is not the case with the drop on the uncoated glass. In this case, the water interacts much more with the glass surface, so friction becomes important. The drop is slowed down by friction as it moves downwards.

23

LIQUID CLIMBERS

Help water climb walls!

SUPPLIES

- 1 shallow dish
- 1 clear cylindrical glass
- 1 clear cylindrical glass jar
- piece of black paper
- paper towel
- water
- water repellent (used on car windshields)

STEP BY STEP

Fill the shallow dish with some water. Put the black paper inside the jar and place the glass and jar in the dish. Press the glass against the outer wall of the jar, as indicated. (Be careful NOT to press too hard, to avoid breaking them.) The more you press (carefully), the higher the water will reach! (The black paper is used for contrast to better visualize the water climbing up.) Now, dry this side of the jar or use the other side. Carefully pour a bit of water repellent on a piece of paper towel and spread it over the strip of the jar surface where the jar will be pressed against the glass. If necessary, repeat this procedure. Wait for a while (read the instructions that come with the water repellent about the necessary waiting time) and then remove the excess water repellent from the side of the jar with a new piece of paper towel. Do the same with the glass. First, press the side of the jar coated with water repellent against an uncoated side of the glass. Is the water still able to climb up? How is the water's performance compared when two uncoated sides are pressed against each other? Next, press the two coated sides against each other. What happens to the water? Can the water grip onto the coated surfaces?

FUN FACTS: COHESION AND ADHESION

The water molecules in liquid water are attracted to each other and stick weakly together, forming temporary chemical bonds. Each molecule has enough "disordered" kinetic energy associated with temperature to break these bonds and then briefly stick to neighboring water molecules before moving on. This binding of the water molecules among themselves is called *cohesion*. In addition, the water molecules adhere to the glass walls of the jar and glass due to attractive forces, called *capillary forces*, which are basically electric forces. (Capillary refers to tubes with a hairlike diameter, which have a huge surface-to-volume ratio.)

FUN FACTS: CLIMBING UP

As the water molecules are more attracted to the glass than to other nearby water molecules, the water rises between the glass walls, forming a peak where the walls are closest to each other. The adhesion of the water molecules to the walls then prevails over cohesion, since a great fraction of them are attracted by the glass of the jar and the glass. The water is only prevented from advancing further up because gravity pulls it down. As the separation between the walls increases, cohesion starts dominating over adhesion (the fraction of water molecules attracted by the glass walls decreases). As cohesion tends to lower surface tension, the peak broadens.

FUN FACTS: WATER REPELLENT

When you apply water repellent to one of the walls, adhesion becomes restricted to the uncoated wall. When both walls are coated with water repellent, cohesion completely dominates over adhesion, and water is prevented from climbing. However, water and water repellent are not as incompatible as it may appear at first sight. Just drop a droplet of water repellent in a container with water. The water repellent will form a very thin film on the water surface instead of remaining as a droplet. This happens because the overall energy balance is more favorable to the film than to the water repellent droplet.

FUN FACTS: PUMPING UP WATER AND WETTING

Capillarity is important for the upward flow of water through very narrow channels in the soil and in plants. In this case, the surface of the channels anchors the water molecules, thus pulling them up. It is also responsible for paper and cloth becoming wet when in contact with water. In paper, the pulp fibers are oriented and woven tightly together. Cloth fibers also are woven tightly together. In both cases, the fibers anchor the water molecules, just as the glass surfaces of the jar and the glass do. Adhesion depends on the surfaces and on the liquid in contact with them. You can repeat the experiment by pressing plastic bottles against each other, glass with a piece of wood, and so on, or else add dishwashing detergent to the water and see how it affects water climbing. To see how beautifully water climbs up paper, make a roll—a cylinder, around 1 1/2 in (4 cm) in diameter and 8 in (20 cm) in height—with a white piece of paper. In a dark room, illuminate the roll with the light beam of a laser pointer hitting its inner wall (you will get diffuse reflection, see Experiment 6, "The Light at the End of the Tunnel" under *Playing with Light: Optics*). Hold the roll vertically with one end in contact with water in a shallow container. Look at the roll's outer surface, opposite to the inner surface, hit by the laser and discover how water wets paper. (This same effect is used in liquid chromatography, important in analytical chemistry.) Enjoy your new findings.

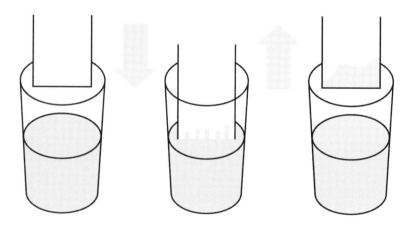

You probably have seen whirlpools in a drain. Find out more about them.

SUPPLIES

- two 2-quart (2-liter) plastic bottles, with caps
- photo film canister or a piece of PVC with inner diameter matching the caps
- drinking straws

STEP BY STEP

Drill or cut a hole in the center of each of the bottle caps, 5/8 to 3/4 in (1.5–1.8 cm) in diameter. Cut the straws into a bunch of small pieces, place them in one of the bottles, and fill it with water. Put the caps on the bottles tightly. Cut the bottom off the film canister and fit it halfway onto the top of the bottle that is full of water. Then, placing the other bottle upside-down over the full one, fit its top into the other side of the canister so that the two bottle tops touch. Carefully invert the positions of the bottles. What happens? Spin the bottle full of water to create a whirlpool vortex.

A STEP FURTHER

Fill the sink in your bathroom with water, move your finger either clockwise or counter-clockwise in the water surface, and then carefully let the water drain out. Observe if the motion of the vortex formed when the water flows down matches the movement of your finger. Repeat the experiment, moving your finger in the opposite direction. Does the water in the sink have memory?

FUN FACTS

When you turn the bottle filled with water upside down, gravity causes the water to flow down through the drain hole connecting the two bottles. Air from the lower bottle forms bubbles and moves up to the surface in the upper bottle to make space available for the water coming in. Eventually, the downward flow will stop because the air pressure in the lower bottle increases relative to that in the upper bottle, and the pressure produced by the water column in the upper bottle decreases. This pressure imbalance together with the surface tension in the water at the drain hole, which pulls the water up, stops the water flow. If you add some dishwashing detergent to the water, more water will flow from the upper bottle to the lower bottle, since the detergent lowers the surface tension. (As the detergent softens the water, it also favors the formation of air bubbles.)

When you turn the filled bottle upside down and spin it around for a while, you will make the water circulate and move down at the same time. A whirlpool with a funnel-like profile is formed. Its free surface is a result of the inertia of the spinning water, which is pressed against the bottle's walls as it circulates (see Experiment 8, "Globe of Death," under *Fun with Mechanics*), and of its surface tension, which tends to minimize the free surface area. As the water moves down, its circulation speed increases. The water circulates at the top along a circle with a bigger radius, and as it goes down, its inner surface comes closer and closer to its spin axis. This means that its rotational inertia decreases as the water approaches the drain hole. To compensate for that, since angular momentum must be conserved, the water circulates faster and faster as it goes down, just as in Experiment 35, "The Ballerina's Trick," under *Fun with Mechanics*.) In the case of the vortex produced in your sink, when you move your finger in a circle in the draining water, the water in the sink acquires angular momentum. The circulation speed thus increases at the drain hole for the same reason it does in the case of the coupled bottles. Watch the vortices formed in your bathtub and discover how the water is drained. (Small objects at the bottom of the bathtub make it easier to visualize the water flow.) Whirlpools are a fascinating and complex phenomenon. There are lots of things yet to be explored using your basic setup for this experiment. How about trying drain holes with different diameters, varying the height of the water columns, seeing how dishwashing detergent affects your tornados, and so on? Can you concoct other ways to better visualize what is going on? Feel free to explore the world of vortices!

OUTLETS CLOGGED WITH WATER

Sometimes even water can become an obstacle for itself.

SUPPLIES

- fine plastic sieve (try different meshes)
- glass jar with plastic screw-on top (you can also use a transparent plastic jar)
- large pitcher, extra large cup, or other suitable clear container with water
- superglue

STEP BY STEP

Remove the top (disk) from the plastic lid to make a frame for the sieve, as shown. Fix the sieve to the frame with superglue and screw it to the jar. Turn the jar upside down and plunge it into the container with water. Remove the jar and, surprise, it is filled with water! Why doesn't the water fall out through the sieve? How coarse can the mesh of the sieve be (how big can its holes be) and still hold the water?

SUPPLIES

- additional jar, identical to the one used in the previous experiment, but with the screw-on top intact
- 2 pieces of Styrofoam, flat and thin, big enough to cover each jar

A STEP FURTHER

Thrill your friends with a cool performance.

STEP BY STEP

Remove the top (disk) from the plastic lid to make a frame, and screw the frame on the jar. In your presentation, you should put the two jars (one with and one without a sieve) side by side so that your audience does not notice the difference. Now, plunge the jar without the sieve upside down into the container of water and cover the opening while it's submerged with one of the pieces of Styrofoam. Hold this cover in place as you remove the jar from the container. The cover will stick to the opening of the jar and the water inside will not come out. Why? Ask an assistant to remove the cover with the water container of the previous experiment placed below it so that the water will fall straight into it. So far, so good. Now do the same with the jar with the sieve. When your assistant removes the Styrofoam cover, no water will fall out this time. You can utter some magic words and tilt the jar a little. At your command, some water will fall out. Then return the jar to its vertical position and no more water will come out. Challenge your friends to explain what is going on. If you notice some impatience, exhibit the two jars. This might, however, not be enough to please your audience.

FUN FACTS

When you submerge the jar without the sieve, it might become partially or totally filled with water, while the air inside comes out to make space available. As you cover the jar with the Styrofoam cover and remove the jar from the water, the atmospheric pressure will be greater than the pressure exerted by the residual air and/or water column inside the jar, thus preventing the water from falling out. As you remove the cover, gravity pulls the water down. When you do the same with the jar with the sieve, the water molecules in contact with the sieve's mesh will become anchored. (Try coarser meshes, larger than a big water droplet, and see if the trick still works.) In addition, the force pulling up due to the molecules on top (surface tension) and the outside air pressure pushing up both help the water to remain in place. Now, if you tilt the jar a bit, the water will slide to the side, allowing air to come in and force water out. When you return the jar to the vertical position, the previous conditions are restored and no more water falls out.

26   FORCING AN EGG OUT OF THE SHELL

See the egg oozing out of the top of a straw stuck in the shell.

SUPPLIES

- a couple of fresh raw eggs (at least 4)
- egg carton
- several egg nests (made from an egg carton, each to hold a single egg)
- drinking straw
- 3 small transparent containers (for example, the lower parts of three 1-pint plastic bottles or 3 glasses)
- fresh water
- superglue or melted candle wax
- piece of solid wire with one pointed end or a needle
- bottle of corn syrup
- toothpicks or pins

STEP BY STEP

A. Watery Eggs

With the pointed end of the solid wire or of a needle, carefully remove a small piece of shell (about the size of your finger nail) from the blunt end of two eggs (there is a small cavity underneath), leaving the inner membrane undamaged. If you break the membrane, start over. Try not to leave sharp jagged edges around the hole. Place the eggs in their nests, surrounding them with plastic or a piece of cloth, so that the open end does not touch the bottom of the nest. Punch a small hole in the sharp end of both eggs with the solid wire (tap it gently with a spoon), enough for the straw to fit tightly. Cut two pieces from the straw, roughly 3 in (7–8 cm) long, and push a piece of straw about one inch into each egg. Drop superglue or melted candle wax around the join to form a seal. Extract the egg white from one of the eggs that is left over. (Punch

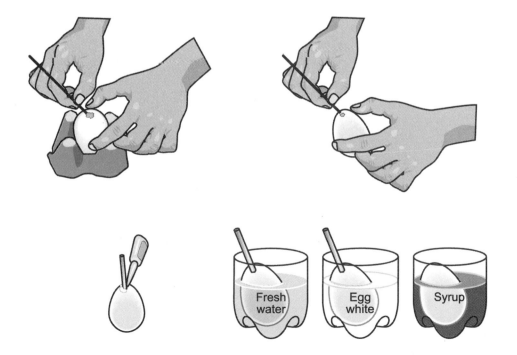

a hole at each end and the egg white will come out.) Partially fill one of the containers with fresh water and the other with egg white. Place the eggs in the containers and leave them. After an hour or two, look and see what happened.

B. Syrupy Eggs

Carefully remove a small piece of shell from the blunt end of an egg, as indicated above, leaving the egg membrane intact. Carefully place the egg in the remaining container and pour syrup into the container, enough to cover the hole in the egg's blunt end and so that the membrane will be in full contact with the syrup. Wait for an hour and see what happens. Try putting the glass of syrup between two crossed polarized filters. Turn one of the polarized filters and see what happens. Is there anything special about syrup?

A STEP FURTHER

STEP BY STEP

Add a drop or two of a strongly scented extract into a deflated balloon. Blow up the balloon, tie it off, and place it inside a plastic bag or a shoebox. To make sure the lid stays on the box, use masking tape to secure it. Also, hold the bag's opening closed with your hand. Lift one end of the shoebox lid or open the bag and smell the contents.

FUN FACTS

The egg's membrane consists of thin layers (actually, it's two membranes, but they are held tightly together) with tiny pores that let small molecules, like water, pass through, but block big molecules, like the sugar molecules in the corn syrup. Membranes that are selectively permeable, like the egg's membrane, are called *semipermeable*. Party balloons also have microscopic pores, which allow vapors of extract to pass through but not the molecules of the liquid ("solute"), which are bound to each other. In the same way that a gas or a liquid moves from a region of higher pressure to a region of lower pressure (partial vacuum), smaller molecules from either side of the membrane move across the membrane until their fraction (partial pressure) is the same on both sides of the membrane. The selective passage of molecules across a semipermeable membrane is called *osmosis*, and the pressure difference produced by the smaller molecules crossing the membrane is called *osmotic pressure*. The egg white is about 90% water. When the external part of the egg's membrane is nearly 100% water, as in A, the water in the container crosses the membrane to balance the water fraction on both sides of the membrane. For the water molecules to move into the eggshell, they need space. The egg is then forced out through the straw to make space for the additional water. When the water fraction is the same on both sides, as in B, nothing noticeable happens. However, when the egg membrane is surrounded by corn syrup, which is about 25% water, things change. In this case, the water molecules move from the side of the membrane where they are more abundant to the side where they are less abundant. So, water migrates from inside the egg to outside the egg. Because water and syrup have different refraction indexes, total reflection is produced as the water content in the syrup increases. (See Experiment 16, "Tubes of Light," under *Playing with Light: Optics*.)

When plant cells take up water by osmosis, they start to swell, but their strong walls prevent them from bursting. The pressure inside the cell rises. Eventually, the internal pressure of the cell is so high that no more water can enter the cell. This liquid or hydrostatic pressure works against osmosis and is what makes the green parts of plants "stand up" in the sunlight. When you forget to water potted plants, you will see their leaves droop. Osmosis is also the basic mechanism used by our kidneys to remove water and salt from our blood, thus keeping their concentration at appropriate levels. Now, it is also possible to force water to cross a semipermeable membrane in the opposite direction of that of osmosis. This is called *reverse osmosis*. To accomplish this, you just have to apply an external pressure. It is used, for example, in water purification and desalination. Can you find a simple way to demonstrate this?

Playing with Sounds:
Acoustics

TELEPHONE WITH A WIRE

Lend a new ear to the old can-and-string telephone.

SUPPLIES

- 2 empty tin cans (like soup cans)
- 11 to 16 yd (10–15 m) of twine
- small piece of cotton or a candle

STEP BY STEP

Make two holes in the bottom of the cans, right in the center. Pull the ends of the twine through the holes and tie knots so the twine can't slip through when it is pulled tight (see the picture). Ask someone to place the can over his or her ear, and stretch the twine while you speak with your mouth very close to the other can. Try also placing the can on your forehead. Can you still hear sounds? Ask someone to hold the can on your forehead or on the back of your head while you keep your ears covered.

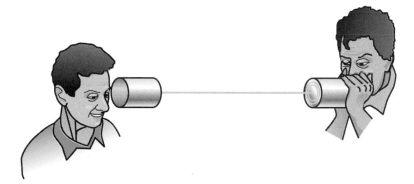

ALTERNATIVES

1. A third person can join the conversation if you have another can and piece of twine. Knot the end of the twine on the inside of the can, pull it through the bottom, and tie it in the middle of the twine connecting the other two cans.

2. You can use just one can. Have someone hold the other end of the twine tightly in his or her teeth or in a circle passed around his or her head or chest. The sound is clearer when the person with the "can-less" end of the twine covers his or her ears.

3. Now, you just use the twine. Each end should be held tightly in a loop around each of your heads. Another alternative is to hold the twine looped tightly around the chest. You should do the experiment outdoors, standing back to back. Both of you should cover your ears. First, keep the twine loose to check if you can hear each other speak. Then, stretch the twine tightly to test if you can hear sound transmitted through the twine. Do you "hear" better with your head or with your chest? Try it both ways: one person with the twine around his or her head and the other with it around his or her chest. Can you find out a way to improve this unconventional way of hearing sounds?

> **HINT**
>
> During the conversation, lightly place your fingers on the twine. Keep the string fixed and see what happens. Find out why the string has to be stretched tightly. (Relax the string and check if you can hear sounds.)

TWO STEPS FURTHER

A. Sizing Up Cans

You can use cans with the same size on each end of the twine, or you can try different sizes. Do the diameter and/or depth affect the acoustic sensitivity on the speaking and listening ends? What about the twine? If you use a thicker twine or a thinner twine, how does it affect the performance of your telephone? Try a mixed twine (for example, a thick piece tied to a lighter piece). How does the stiffness or flexibility of the bottoms of the cans affect the transmission? What would happen if the bottoms of the cans were stretched pieces of balloon?

B. Cotton Mouth

Hold the palm of your hand in front of your mouth. Does the air pressure vary while you speak? Now place a small ball of cotton on a table. With your mouth at the same level as the ball and about 4 in (10 cm) from it, spell some words (in a low tone). Does the ball dance while you speak? Do different letters or syllables make it dance differently?

Also, with an adult present, speak close to the flame of a candle. Be careful not to burn yourself! (Use a tube of paper with one end placed about 2 in (5 cm) from the flame and keep your mouth close to the other end of the tube.) Does the flame dance while you speak? Again, you can compare the dances to the different letters or syllables you say. Also, use the palm of your hand to feel your throat, the top of your head, your neck, and your chest while you speak. Do other parts of your body also vibrate when you produce sounds?

FUN FACTS

As you speak, the air pressure fluctuates up and down around normal atmospheric pressure, causing the back and forth dancing of the cotton ball and the candle flame (see also Experiment 11, "From Lungs to Mouth," in this part of the book). These compressions and rarefactions propagate due to collisions between air molecules. The waves of pressure you produce while speaking make the bottom of your can vibrate like a drum as the air pressure increases and decreases. (The same happens inside your ear as the little bangs of air molecules against the eardrum add up to further transmit the sounds you hear. It is just the opposite of what happens when you play a drum—as you beat the drum, you produce sounds; here sound makes the drum vibrate.) The twine end attached to the bottom of the other can is either pulled or relaxed as the twine becomes more or less stretched. The air pressure inside the can increases or decreases as the bottom advances or recedes. The bottom vibrates like a drum, producing sound. The dancing of the bottom of both cans just follows the rhythm of your voice. Besides propagating in air, sound can also propagate through our skull and reach the eardrum, as the experiment with a single can demonstrates. This is called bone conduction of sound. What about hearing two simultaneous calls through this alternative sound conduction mode?

2

SCRATCHING MADE LOUDER

Make scratching louder with just a piece of paper!

SUPPLIES

- piece of solid wire (try different lengths)
- piece of paper
- piece of plywood, rigid plastic, metal sheet, or a stone (why not?)

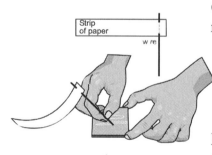

Piece of solid wire

Piece of wood

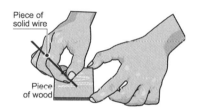

Strip of paper

w re

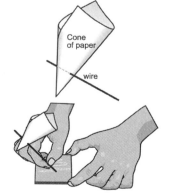

Cone of paper

wire

STEP BY STEP

Scratch the surface of the plywood, for example, with the wire. Can you hear something? Now, fasten a strip of paper to the wire, like a flag. Scratch again and compare the sound with the sound from the wire alone. Finally, make a funnel out of a piece of paper and fasten the wire to it, as shown. Scratch as before and hear the sound that comes out from the opening of the funnel. In which case can you produce the loudest sound?

A STEP FURTHER

Cup your hands (fingers tight together) to make two shells. Place your cupped hands behind your ears with the thumbs close to your head, as if your hands were an extension of your ears. Do you notice any difference in the sounds you hear (your own voice and other people's voices)? Reverse your hands so that your knuckles are now in contact with your head at the opposite side of your ears. What happens to the sounds you hear?

FUN FACTS

If you scratch the wood surface with just the piece of wire alone, the vibrations produced in the wire are transmitted to the air around

Playing with Sounds: Acoustics

it, producing sound. Since the wire surface has a very small area, you hear a very low-volume sound. When you fasten a strip of paper to the wire, the vibrations produced in the wire are transmitted to the strip, which has a larger area, so the sound is amplified. The funnel, in addition to having a still larger area, is more efficient than other geometries for sending the sounds out into the environment, just like a loudspeaker. (Replace the funnel, for example, with a corner made of paper, and try other shapes, using pieces of paper with the same size as that used in making the funnel.) The funnel works like the intermediary ball in A Step Further of Experiment 17, "Bouncing Balls," under *Fun with Mechanics*. This ball (funnel) allows a more efficient transfer of energy (sound) from the bigger ball (the source of sound) to the smaller ball (the environment). Engineers and physicists call this cute effect *impedance matching*. It is also present in ultrasound imaging. Between the probe (the ultrasound source and receiver) and the body being scanned there is an air gap. Since the speed of sound in air is much less than inside the human body, the sound emitted by the probe is strongly reflected back at the skin. To avoid that reflection, a gel is used between the ultrasound probe and the skin to eliminate the air gap. The ultrasound can now propagate from the source without reflection. So, you match the sound source with its environment. The cone of paper works in a similar way. When you place your cupped hands close to your ears, you might also get a better acoustic match between your ears and the environment for certain frequencies. Now, what about replacing the piece of paper with construction paper or plastic? Does it produce louder sounds? What happens then?

3

WHEN IS A PIPE A BELL?

Give a pipe a chance to become a bell!

SUPPLIES

- piece of thick wire or metal pipe approximately 1.6 ft (50 cm) long (could be a thin piece of iron, not necessarily hollow, but not heavy)
- 2 pieces of string of equal length, 1.6 ft (50 cm) each
- 2 plastic bottles (1 pint or 500 ml) with caps, or two plastic cups plus a stapler

STEP BY STEP

Cut the two bottles about 2 3/4 in (7 cm) below the opening, making two funnels (headphones). Make a hole in the center of each bottle cap, just wide enough to pass the string through, from the outside in. Then knot the ends inside the caps. Screw the caps on the funnels, which are now the headphones, and tie the other ends of the strings around the pipe, as the picture shows. The headphones could also be made of disposable cups. In this case, just staple the string to the side of the cup (the staple makes the side stronger so it won't tear). Now place the headphones against your ears and ask someone to "play" the wire or pipe with a screwdriver. Don't let the pipe touch your body. Let it hang freely in the air, suspended by the two strings. What do you hear?

FUN FACTS

When the wire or pipe is hit, it vibrates. The vibrations propagate along the strings, making them stretch or relax, just as with cans in Experiment 1, "Telephone with a Wire." (If you ask someone to hold the two strings fixed in the middle at the same time, you will hear no sound.) The bottom of the headphones thus follows the rhythm imposed by the vibrating wire, producing sound. You could press the string right next to your ear and produce a similar effect. What are the advantages of using cups? Do they produce a higher volume? Why?

Listen to the voice of a wind-up watch.

STEP BY STEP

Place the watch the same distance away from you as the length of the tube. Can you hear something? Place your ear against the opening of the tube. What do you hear? Now, place the watch inside the tube, near the end away from your ear, and close the end off with your palm. What can you hear now? What does the tube do, after all? What relationship can you find between this experiment and seashells held up to your ear?

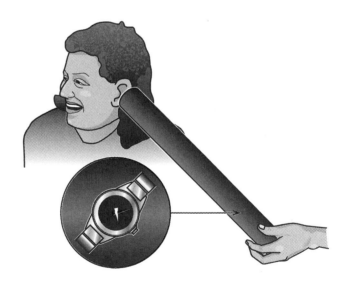

A STEP FURTHER

A. Tick-Tock in a Funnel

You can also use the corrugated hose (see Experiment 10, "Singing Hose," in this part

of the book) with a funnel made from a plastic bottle at one end. Place the watch in the funnel and hold your palm over the end, making a seashell shape. Place the other end of the hose over your ear and have fun with the tick-tock of the watch. So, now, can you say, does sound curve?

B. Funnel Your Heartbeat

Cut a piece from a party balloon, stretch it until it covers the wide end of the funnel, and use the masking tape to attach it to one end of the hose. The other end should have another funnel. This is your homemade stethoscope! Press the funnel covered with the balloon to your chest, over the heart, and hold the other funnel over your ear. Now you can listen to your heartbeat.

FUN FACTS

A flat surface can act like a flat mirror for acoustic waves (sound), much in the same way as flat mirrors reflect light (see Experiment 6, "The Light at the End of the Tunnel," and Experiment 7, "The Ghost Behind the Mirror," both under *Playing with Light: Optics*), as the experiment demonstrates. This happens because sound propagating in air involves collisions of air molecules. When the molecules hit the surface, they are reflected like billiard balls. The overall effect is the sound reflected like light in a flat mirror. In the case of sound propagating inside a hose, the hose's inner walls reflect the air molecules, guiding the molecules and, as consequence, sound itself.

5

WIRELESS TELEPHONE: PARABOLIC ACOUSTIC MIRRORS

Attune yourself and communicate!

SUPPLIES

- plastic ruler (flexible) with measurements, 1 ft (30.5 cm) long
- wooden disk, 3/8 to 5/8 in (1 to 1.5 cm) thick and about 2.3 ft (70 cm) in diameter
- other wooden pieces, 5/8 in (1.5 cm) thick
- thick (stiff) posterboard or cardboard, preferably colored
- paper glue or superglue

STEP BY STEP

Draw a rectangle 1.3 x 2.5 ft (40 x 75 cm) on a piece of posterboard. Successively mark the points P_1, P_2, P_3, P_4, P_5, P_6, and P_7, referring to the measurement table and the X-Y graph shown in the figure. Bend the ruler so that it passes through points P_1, P_2, and P_3, and with a pen or pencil, continue the curve defined by the ruler (a parabola). Do the same with points P_3, P_4, and P_5, and finally with P_5, P_6, and P_7. As the X-Y graph shows, the tiny brown square should be discarded. (In the center of the disk, the points of all the pieces can't meet because the wood has its own thickness.) Cut out the pieces, as shown in the illustration, giving you the eight pieces necessary for the structure of a parabolic mirror. Try to make the best use possible of the available material, avoiding waste. Divide the disk into eight equal parts (see design). Attach each of the eight wooden pieces you cut from the pattern to the lines on the disk, so that the ends all meet in a circle of 3/4 in

MEASUREMENT TABLE			
X (ft)	Y (in)	X (cm)	Y (cm)
$X_1 = 0$	$Y_1 = 1\ 1/4$	$X_1 = 0$	$Y_1 = 3$
$X_2 = 0.5$	$Y_2 = 1\ 5/8$	$X_2 = 15.6$	$Y_2 = 4$
$X_3 = 0.8$	$Y_3 = 2\ 3/16$	$X_3 = 24.7$	$Y_3 = 5.5$
$X_4 = 1.3$	$Y_4 = 3\ 3/4$	$X_4 = 39.8$	$Y_4 = 9.5$
$X_5 = 1.8$	$Y_5 = 6\ 1/8$	$X_5 = 54.3$	$Y_5 = 15.1$
$X_6 = 2.0$	$Y_6 = 7\ 1/4$	$X_6 = 61.1$	$Y_6 = 18.3$
$X_7 = 2.5$	$Y_7 = 10\ 5/8$	$X_7 = 76.2$	$Y_7 = 26.8$

(2 cm) radius (draw this in with a compass). It will help a lot if you keep all the angles equal.

If necessary, reinforce the structure of the mirror with wooden strips, and build a base to hold up the mirror. To cover each of these eight sections of the mirror, place the poster board on the structure and draw for one section the corresponding area where the wood will touch (see picture). If you made all the angles equal, you can use that as a pattern for cutting the eight pieces. Use the ordinary glue or superglue to attach all the poster board pieces to the wooden supports. If necessary, use a few small nails. As a stand for the acoustic mirror, use a wooden column 1 yd (roughly 1 m) high, attached to a base and attached with screws to the back of the mirror (see illustration).

The Focal Point

The mirror is dimensioned to give it a focal point 2 ft (61 cm) from its center (see illustration). It would be helpful to place a wire circle, identifying the focal point, held in place by three straight wires from equally spaced points on the edge of the mirror.

Wireless Telephone

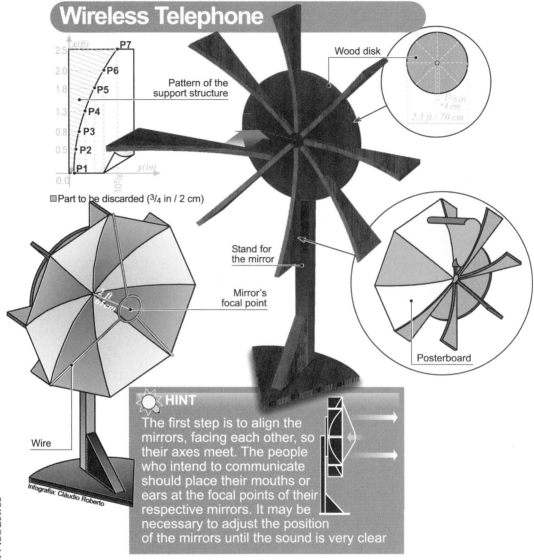

x(ft)

2.5 — P7
2.0 — P6
1.8 — P5
1.3 — P4
0.8 — P3
0.5 — P2
0.0 — P1

y(in)

Part to be discarded (³/₄ in / 2 cm)

Pattern of the support structure

Wood disk

1⁵/₈ in / 4 cm
2.3 ft / 70 cm

Stand for the mirror

Mirror's focal point

2 ft / 61 cm

Posterboard

Wire

Infografia: Cláudio Roberto

HINT

The first step is to align the mirrors, facing each other, so their axes meet. The people who intend to communicate should place their mouths or ears at the focal points of their respective mirrors. It may be necessary to adjust the position of the mirrors until the sound is very clear

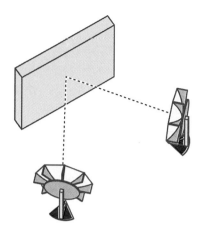

ALTERNATIVE

Try communicating with someone by means of these acoustic mirrors on a rainy day, with their axes aligned, so that the sound will propagate in open air. To avoid the mirrors getting wet, keep them under a shelter. What effect does the rain have?

TWO STEPS FURTHER

A. Sound Reflection

Place the mirrors as shown in the illustration to demonstrate sound reflection off walls. Actually, you can use of the setup you built for Experiment 6, "The Light at the End of the Tunnel" (under *Playing with Light: Optics*) and place a wind-up watch at the end of the immovable tube. Vary the angle between the tubes, keeping your ear over the end of the movable tube. (Also, see Experiment 4, "Tick-Tock of the Clock" in this part of the book.)

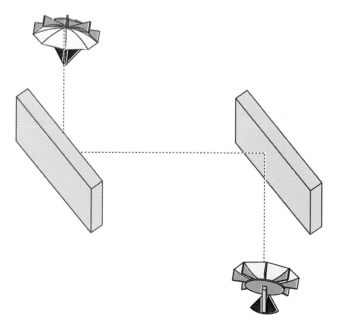

B. Acoustic Periscope

Place the parabolic mirrors so that their axes are parallel and place two vertical wooden boards 1.6 x 2.2 ft (1.5 x 2 m) as flat acoustic mirrors, as shown, so that they each form a 45-degree angle to the axis of each parabolic mirror.

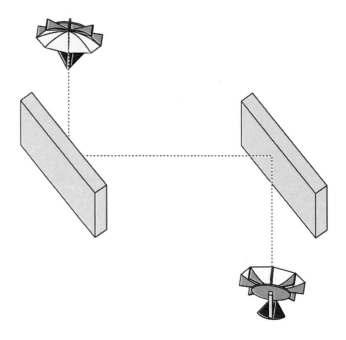

> **HINT**
>
> Use a drafting square with a 45-degree angle in order to correctly align the wooden boards. You can also use a laser pointer and two flat mirrors to obtain the periscope's condition.

FUN FACTS

A parabolic sound-collecting dish is essentially a parabolic antenna for sound. You might have already noticed that car headlights, like pocket flashlights, have parabolic mirrors surrounding the bulb inside. This makes the light directional. If parallel light rays hit the inside of the mirror along its axis, after reflecting, they all concentrate at a point called the focus of the mirror. Satellite and giant radiotelescope antennas also work in the same way. The same holds for a parabolic sound-collecting dish. That means all of the incoming sound waves bounce off the dish and converge toward the focus, synchronized so that they work together to make the loudest possible sound vibrations. The sound is thus enhanced at the focus, but only if it originated from the source you're aiming at. Sounds from other sources miss the focus. If the direction of propagation of the collected sound waves is tilted in relation to the antenna's axis, the sound is enhanced at a different point in the focal plane of the mirror (a plane perpendicular to the antenna's central axis containing its focus). The antenna also works in reverse. If you emit sound at the focus, the sound waves will bounce off the mirror and propagate parallel to the antenna's central axis, just as in flashlights. (It is just a reverse playback of the directional incoming sound waves.) Small antennas scatter away sounds with wavelengths comparable to the antenna's dimensions. This means that only the higher frequency sound components are focused. This is analogous to the scattering (diffraction) of light when it hits obstacles with dimensions of the same order of the light's wavelength. (See also Experiment 13, "Car Control versus TV Control," under *Electrifying Experiments: Electricity and Magnetism*.) So, in order for the parabolic antennas to work well as sound mirrors, they must be big enough to focus all the desired frequencies. Also, their walls must be sufficiently rigid (otherwise the sound would make the mirror and support structure itself vibrate, and the sound would be lost).

6

FOCUSING SOUND

Can you focus sound as you can light?

- party balloons of different sizes
- 1-gallon soda bottle (filled with soda) or 1-pint empty plastic bottle, vinegar, and baking soda

STEP BY STEP

Open the soda bottle and quickly slip the mouth of a balloon over the bottle top. Move the bottle in circles to speed up the process, as indicated in the figure. Alternatively, you can put about 8 ounces of vinegar in an empty soda bottle. Add about a tablespoon of baking soda. (In the case of bigger balloons, you can increase the amount of these components, keeping the same proportion. Also try the following: after adding the vinegar, pour in enough water to nearly fill the bottle. That causes more of the CO_2 to go into the balloon.) Quickly slip the mouth of a balloon over the bottle top. When the balloon is full, tie it closed. Also, have a balloon filled with ordinary air for comparison. In a very quiet room, hold the balloon between your ear and some source of low-volume sound, such as a ticking watch. Move the balloon nearer and farther away from your ear until the sound is loudest. How does it compare with the air-filled balloon? Try balloons with different sizes and shapes, and compare their performances. You can also put your balloon aligned with the axis of a parabolic acoustic mirror (see Experiment 5, "Wireless Telephone") at some distance from it, let's say between 50 and 100 yards, and ask someone to talk to you. Compare the sounds you hear with and without the balloon.

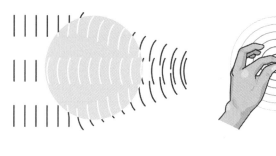

A STEP FURTHER

If you put your balloon filled with CO_2 or with air in the freezer for long enough, the temperature and volume of the balloon will both drop. This increases its density. (See this experiment's Fun Facts.) Does making it much colder decrease its density enough so that you can notice a difference in its performance as a sound lens?

FUN FACTS

The sound of a ticking watch, for example, travels into the balloon. Since the balloon is filled with carbon dioxide, a gas that is more sluggish than air, sound travels at a lower speed in CO_2 than it does in air. (The collisions between molecules are ultimately responsible by the propagation of sound, so the greater their mass, the more slowly sound propagates.) This implies that sound changes direction when it passes through the balloon, much as light bends when it passes through a lens, as shown in the figure. Imagine a line of people walking at an angle to the shoreline.

As each person enters the water, he or she is slowed down and the line is bent. The people in the water are then facing in a different direction and therefore travel in that new direction. Similarly, the direction of travel of the sound waves changes as they enter the CO_2. The same holds for light rays entering an ordinary lens. The balloon actually acts like a lens, since both have a curvature and a refraction index different from air (the speed of propagation of light is lower inside the lens than in air, just as sound propagates more slowly inside the balloon filled with CO_2 than in air). Sound is then focused near your ear as if the sound source were right next to you. What happens to the components of sound with wavelengths longer than the balloon's radius (lower frequencies)? Can you use a ball filled with air as a sound lens under water? Does it focus the sound produced when you move your hand quickly back and forth under water?

HINT

The CO_2 is what you see foaming and bubbling when you mix vinegar and baking soda or when you open a soda bottle or a bottle of carbonated water. The CO_2 is injected into the bottle at a very high pressure and dissolves in the liquid. When you open the bottle, the outside atmospheric pressure is much lower and the CO_2 comes out. That's why you can fill the balloon by slipping its mouth over the bottle top.

7

HOME-MADE VARIABLE-PITCH WHISTLE

Have you ever thought of making your own whistle? Here is your big chance to make it happen!

SUPPLIES

- plastic tube from a pen, without the ink cartridge
- wood stick
- piece of rigid rubber
- piece of wood or metal
- superglue

STEP BY STEP

Make an opening and a slot in the plastic tube, as shown. (You can make several models by using different positions and dimensions for the opening and slot.) Glue a piece of thin wood or metal sheet into the slot. Now glue a round piece of rubber on the end of the stick so that when you insert it in the tube, it slides tightly (but easily). Just blow at the left end of the plastic tube and move the wooden stick as shown to produce different sounds.

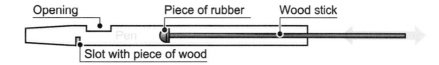

FUN FACTS

When you blow into the plastic tube, turbulence is generated as the air flow is forced through the constriction in the plastic tube. The air pressure then fluctuates up and down, producing sound waves. The cavity between the constriction and the rubber end of the stick selects which frequencies dominate in the sound emitted by the whistle. More than one mode can vibrate simultaneously when you blow across the constriction. The air is then easily deflected and tends to follow the rhythm of the air inside the cavity as it bounces back and forth between the fixed rubber end and the constriction, where the impedance mismatch is high (see Experiment 2, "Scratching Made Louder"). The strengthening of the vibration inside the cavity is similar to the resonant energy transfer that occurs when you make a pendulum swing (see Experiment 18, "Temperamental Pendulums," under *Fun with Mechanics*). The shorter the length of the whistle's cavity, the higher the frequencies of its vibrational modes. Hold a flexible plastic ruler flat on a table top with one end (the free end) over the edge of the table. Now make the free end vibrate by "plucking" it. As you shorten the vibrating part of the ruler, its mass decreases and the stiffness of the restoring force becomes greater. Both factors lead to a higher frequency of vibration. The same holds for the air in shorter cavities. You can also produce sounds with selected frequencies when you blow into an empty soda bottle. (In this case, you can fill the bottle with water to change the air volume and the cavity's resonance frequencies, as you do by moving the rubber end of the whistle.) Now, how can you tune up two variable-pitch whistles? What happens if they are blown simultaneously and one of the wood sticks or rubber ends is slightly displaced? What about producing an improved version to get sharper sounds?

Paper, too, has something to say.

Playing with Sounds: Acoustics

SUPPLIES

- pieces of ordinary paper or newspaper
- shallow container with water, big enough for soaking sheets of paper

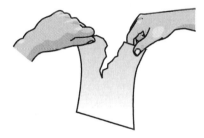

STEP BY STEP

Tear pieces of paper along different directions with different sizes and tearing speeds. What sounds do you hear? Now, wet the paper (let it soak for a while) and tear it again. Do you still hear sound at all? Try crumpling the paper. Try different crumpling speeds. You can stretch out the sheet after you have made a compact ball out of it and crumple it again. Does it make any difference for the sound produced by crumpling? Is the paper size important for the sound you produce by tearing and crumpling paper? You can also explore the friction between two paper sheets by sliding one past the other. Does it produce a distinct sound?

FUN FACTS

As you tear paper, there is a localized rapid release of elastic energy at the tear edge with the production of sound pulses, which you can consider impulsive events. Think of stretching a piece of thread or elastic apart to the point of rupture. A distinct sound pulse is produced. (You can fix both ends of the stretched elastic and heat up the middle of it with the flame of a candle—you will hear a distinct sound when rupture occurs at the spot where the elastic is heated.) In paper, you can have a rupture of a bond, of a fiber, or of both. As you tear the paper faster and faster, the number of ruptures per second increases and the rate of sound pulses also increases. Eventually, they coalesce into a continuum of sounds.

It turns out that many of the systems that physicists are interested in, such as earthquakes, magnets, and noise pulses emitted by materials under stress, emit pulses of energy similar to the way paper does when you tear or crumple it. Playing around with paper puts a world of thrilling science in your hands.

9   SECRETS OF THE GUITAR

Find out what each piece of a guitar is for!

SUPPLIES

- clear plastic box
- rubber band
- TV screen or computer monitor
- shell of a sea snail

STEP BY STEP

Stretch the rubber band and pluck it with your finger. Build a primitive guitar by stretching the rubber band around the box (without a lid), as the picture shows. Compare the sound you heard from the rubber band alone with the sound from your guitar box. What is the body (box) of a guitar for? Stretch the rubber band a bit more each time, keeping it in contact with the sides of the box, and pluck the rubber band with your finger to make it vibrate. What are the tuning pegs of a guitar for? Now place your guitar between you and a TV screen or computer monitor (see Experiment 17, "Slow Motion Camera," under *Playing with Light: Optics*). What can you see when the rubber band vibrates?

A STEP FURTHER

Fill the box gradually with water and repeat the experiment. What happens when the volume of air in the box is decreased? Now fill the snail shell with water. Can you still hear the "roar of the ocean"?

FUN FACTS

As you pluck the strings of a guitar or any other stringed instruments, they vibrate up and down, sweeping the air around them and producing sound. Since new transversal waves are continuously generated as you pluck the strings, only those superposing constructively survive. (Waves with other wavelengths superpose destructively and hence do not produce sound.) The strings thus select the frequencies of the sound generated, which correspond to the resonance frequencies of the strings (see Experiment 18, "Temperamental Pendulums," under *Fun with Mechanics*). In addition to the string length, the selected frequencies depend on the string mass and on how stretched it is. The volume of the sound produced by a vibrating string is still low because the string area is too small (see Experiment 2, "Scratching Made Louder," in this part of the book). The vibrations produced, however, propagate to the "sound boxes" of the instruments, which have a much larger contact area with air (see the two previous experiments). The motion of the belly of the instruments in and out thus produces a considerable amplification of the sound. Some of this sound comes from the belly's outer surface, while some comes from its inner surface and must emerge through the opening in the instrument's belly. When the resonance frequencies of two stretched strings placed side by side coincide, you can make one of them vibrate as you pluck the other. The sound you generate when you pluck one string (rubber band) has the same frequency as the oscillating string. This sound, in turn, makes the other string (rubber band) vibrate as the sound's energy transfer is synchronized with the vibration rhythm of the string. (A vibrating string produces sound, and sound makes a string vibrate!)

10 SINGING HOSE

Come alive to the music of corrugated hoses!

SUPPLIES

- corrugated hose (ridged hose) 2 to 2.6 ft (60–80 cm) in length

STEP BY STEP

You just have to hold one end of the hose and swing it around, without closing off either end. You'll hear very nice sounds that change as you spin the hose at faster speeds. You can also use the corrugated hose with a funnel made from a plastic bottle at one end (see figure).

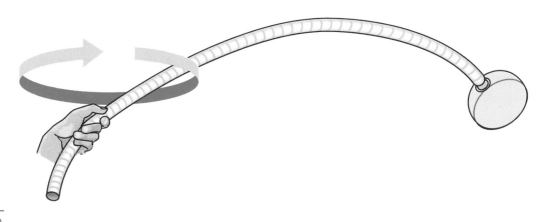

SUPPLIES

- piece of transparent (smooth) plastic hose of 2 to 2.6 ft (60-80 cm) in length
- 2-quart (2-liter) plastic bottle
- bucket
- water
- food coloring (for contrast)
- masking tape
- party balloons

1. Fill the bottom part of the bottle with water mixed with food coloring. Ask someone to hold one end of the transparent plastic hose immersed in the water while you swing the hose as indicated. Keep an eye on the water column formed close to the submerged end of the hose while you swing the other end. What happens and why?

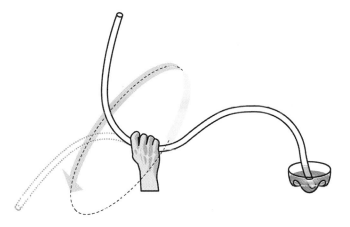

2. Cut the plastic bottle near the bottom to make a long cup out of it. Screw one of the ends of the corrugated hose onto the bottleneck. (If necessary, use masking tape to make the fitting tight.) Now fill the bucket with water and plunge the cup into the water; then move it up. Keep moving the cup up and down to hear the hose singing again and again. Where does the hose's "voice" come from? What's special about the corrugated hose? Try the smooth hose. Does the trick still work?

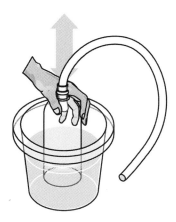

3. Fill a party balloon with air and then stretch its neck, as shown, letting air come out. Why is a sound (not as nice as the singing hose) produced? What about making a duet with two balloons? (Ask someone to help you.) If you stretch the balloons differently, you might hear sound beats (a distinct sound, different from the sounds produced by each balloon separately).

FUN FACTS

The air inside the hose is submitted to different radial forces as it moves in circles with different radii. It is similar to Newton's bucket (see Experiment 9, "Flattening the Earth at the Poles" under *Fun with Mechanics*), except that now there is no wall to prevent the air from escaping. At the spinning end, the air pressure becomes lower than atmospheric pressure, producing a pressure imbalance, as step 1 allows you to demonstrate. The air at the fixed end is sucked toward the spinning end, generating an air flow inside the hose. In step 2, as you move the cup down, the volume available for the air inside the cup decreases and the air pressure increases, so you force the air out through the hose. When you move the cup up, the volume available for air inside it increases and the air pressure decreases. As a result, the air is sucked in. However, the mere flow of air across the opening of a hose is not enough to produce sound, as a smooth hose demonstrates. The corrugations play a vital role in producing the sound. To see how sound is produced, let's first look at the balloon filled with air and with stretched necks. Inside the neck, the pressure energy of the air is converted into kinetic energy, so the pressure inside the neck decreases as the air comes out. As a result, the air outside "chokes" the balloon. The air flow through the balloon's neck then decreases, while at the same time the pressure inside it keeps forcing the air out. This intermittent flow of air with rapidly varying air pressure produces the sound you hear. If you put everything together, you can understand why the corrugated hose is able to modulate the air passing through it and make the hose sing.

11   FROM LUNGS TO MOUTH

Speech sounds spellbound!

SUPPLIES

- party balloon
- piece of PVC pipe 6 3/4 in (17 cm) in length and inner diameter of roughly 1 1/4 in (3 cm)
- piece of rigid wire 1.5 ft (about 45 cm) in length
- cork with a diameter smaller than the internal diameter of the PVC pipe
- plastic bottleneck
- masking tape

STEP BY STEP

Get one end of the rigid wire stuck into the cork (this represents your tongue). Fasten the bottleneck to one end of the PVC pipe. (You may have to make several cuts in the bottleneck perpendicular to its edge to fasten it more easily to the PVC pipe, using masking tape.) Fill the balloon with air and twist its neck so that air doesn't escape. Now, ask someone to insert the "tongue" into the PVC pipe (your "vocal tract") and slip the mouth of a balloon over the bottle top. Untwist the balloon's neck and stretch it to produce sounds while the tongue is displaced along the PVC pipe. Can you notice different sounds as the tongue is displaced? If you assign to the stretched neck the role of your vocal folds and to the filled balloon the role of your lungs, then you will be prepared to follow how vowels are produced.

Playing with Sounds: Acoustics

293

A STEP FURTHER

Try different PVC pipes with different lengths (vocal tracts with different lengths) and corks (tongues) with various sizes, and stretch the balloon's neck in different ways. Now, how about using a corrugated hose as a vocal tract? Can you make your model sing?

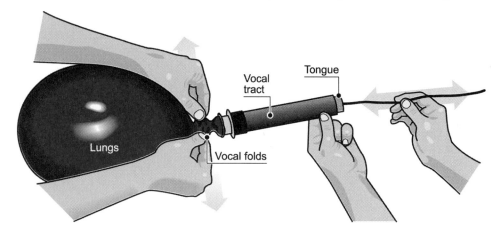

Lungs
Vocal folds
Vocal tract
Tongue

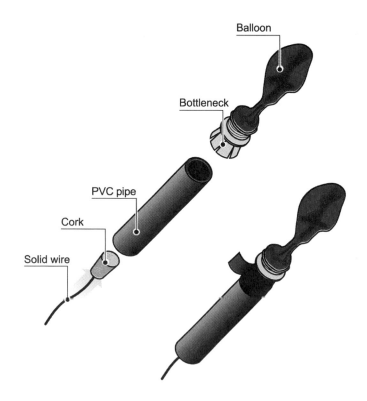

Balloon

Bottleneck

PVC pipe

Cork

Solid wire

As you stretch the balloon's neck, you produce sound (see step 3 of previous experiment and the corresponding Fun Facts). In this model, the filled balloon corresponds to your lungs. The air flow through the glottis (the space between the vocal folds, represented by the stretched necks) is modulated by the movement of the vocal folds, producing a modulated air pressure. The sounds thus produced are shaped into the various sounds of speech in the mouth cavity, where they are selectively amplified as you move, for example, your tongue (represented by the cork), lips, and jaw. As you displace the rigid wire stuck to the cork in and out, you simulate some of these possibilities. Explore them all.

---

12   PICTURES OF SOUNDS

Transform sounds into cool pictures!

SUPPLIES

- party balloon
- big empty can (for example, one that held fruit or vegetables)
- flexible plastic hose, like a corrugated pipe
- plastic bottle
- fine sawdust or chalk dust
- small lightweight piece of a mirror
- laser pointer
- clothespin
- masking tape or a rubber band
- 5 long, thin pieces of wood

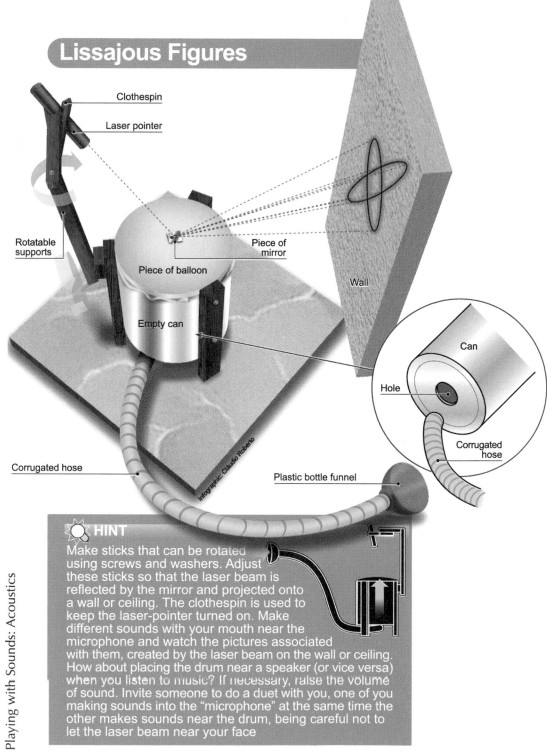

Lissajous Figures

Clothespin

Laser pointer

Rotatable supports

Piece of balloon

Piece of mirror

Wall

Empty can

Can

Hole

Corrugated hose

Corrugated hose

Plastic bottle funnel

Infographic Cláudio Roberto

HINT

Make sticks that can be rotated using screws and washers. Adjust these sticks so that the laser beam is reflected by the mirror and projected onto a wall or ceiling. The clothespin is used to keep the laser-pointer turned on. Make different sounds with your mouth near the microphone and watch the pictures associated with them, created by the laser beam on the wall or ceiling. How about placing the drum near a speaker (or vice versa) when you listen to music? If necessary, raise the volume of sound. Invite someone to do a duet with you, one of you making sounds into the "microphone" at the same time the other makes sounds near the drum, being careful not to let the laser beam near your face

Make a hole in the bottom of the can, exactly in the center, to fit the hose. Screw three of the wooden pieces to the sides of the can, as shown. They are legs to hold the can up high enough to easily accommodate the hose. Cut the bottle top, making a funnel shape ("microphone") and place it inside the end of the hose (see Experiment 10, "Singing Hose"). Cut the balloon open and stretch a piece of it over the top of the can, securing it with the rubber band or masking tape, making it a kind of small drum. Spread the sawdust over the top of the drum. Watch the "picture" your voice makes on the top of the drum when you make different sounds into the microphone. Now, how about combining sound and light?

You can produce similar visual effects using the setup of part A in Experiment 15, "Exploring the Laser Beam," under *Playing with Light: Optics*. The laser beam should cross the water just beneath its surface and be kept parallel to it. If you produce waves on the water surface with, for example, a finger or a spoon, you will see the light spot of the laser beam projected on a wall ("screen") dancing quickly around. You can also superpose different waves and see how the laser beam reacts. Try making the laser beam cross the water at different depths and see what happens to the projected light spot when you produce waves on the water surface. Enjoy yourself with your new findings!

FUN FACTS

As you produce sounds in the "microphone, the air pressure fluctuates around normal atmospheric pressure inside the can. As a result the stretched piece of balloon vibrates. Since it is an elastic membrane, it moves up and down in a more pronounced way than the bottom of a can would do (see Experiment 1, "Telephone with a Wire," on page 267.) The small mirror attached to the balloon thus also vibrates. Since the laser beam will be reflected to a different point for each position of the mirror, the spot of light created by the reflected laser beam on a "screen" (wall or ceiling) will follow the rhythm of the sounds you produce. The reflected spot of the laser beam moves along the *arc length* of the changing angle, which is the change in angle (in radians) times the distance from the mirror to the screen. So, the more distant the mirror is from the screen, the bigger the visual effect produced. If you have a sufficiently lightweight mirror and a laser *(always observe safety precautions when using lasers)*, you can stick the mirror to a person's wrist with a dab of glue and use the reflected laser beam to read the pulse. This same idea is also used in a wide variety of industrial and research applications to magnify and measure small motions. Can you use it to "see" your heartbeat or how other parts of your body vibrate when you speak?

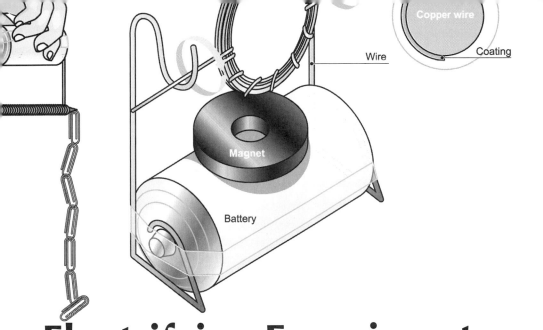

Copper wire

Wire

Coating

Magnet

Battery

Electrifying Experiments:
Electricity and Magnetism

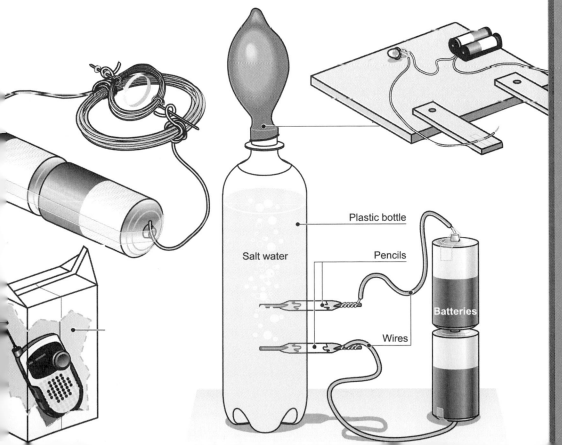

Plastic bottle

Salt water

Pencils

Batteries

Wires

STICKING BALLOONS ON WALLS: STATIC ELECTRICITY

Steal the show using only balloons and your head.

STEP BY STEP

Fill up the balloon and tie a knot in the end to keep the air in. Rub the balloon on your hair, which must be clean and dry, and then place it on the wall or ceiling. What happens?

HINT

On humid days the trick doesn't work (as well). Try getting the balloon wet and watch the effects.

FUN FACTS: SLIDING FRICTION AND STATIC ELECTRICITY

Electric charges are present in all objects and are responsible for holding together all kinds of matter consisting of atoms. Usually, the objects around us have equal amounts of negative electric charge (electrons) and positive electric charge (protons), so there is no net electric charge in them. You produce static electricity, for example, when you rub two surfaces (sliding friction). If one of the surfaces attracts electrons from the other surface, it gains an excess of negative charge relative to the uncharged state. If the surface is nonconducting (an insulator), additional electrons tend to remain relatively static, trapped on the surface until the surface comes in contact with something metallic, where the electrons are mobile. The same reasoning holds true for the surface willing to give up electrons. It becomes positively charged due to its deficiency of electrons relative to the uncharged state. Rubber, plastic, vinyl (PVC), and Scotch tape, for example, are very greedy and always try to steal electrons from surfaces they come in contact with. Pull Scotch tape off a surface, and it will become charged. Dry skin, hair, and glass, for example, are more willing to give up electrons. Static electricity is thus produced when you rub a rubber balloon on your hair or rub soft dry plastic on a glass. This process is generally called *charging by friction.* It also happens when you walk on a carpet. As you walk, sliding friction transfers negative charge from the carpet to the highly insulating rubber or plastic soles of your shoes. (If you have leather soles, it is the other way around.) When you sit in a chair, the contact between your clothes and the chair can also generate a lot of electrostatic charge on your clothes through sliding friction.

FUN FACTS: ELECTRIC FORCES

Charged bodies can attract or repel each other, depending on whether they have net charges with different signs (attraction) or the same sign (repulsion). Rub a balloon (filled with air) on your hair and ask someone else to do the same with another balloon. Watch what happens to the other person's hair and to your own "flyaway hair." Then bring the two rubbed parts of the balloons close to each other, holding the balloons by a string attached to their necks, and see what happens. Now, move the charged balloon close to someone's charged hair. You can also rub a glass on soft, dry plastic (you can use, for example, a plastic handbag) and see how the charged balloon reacts towards it. Now, when you bring the charged balloon up to the neutral wall, the charge on the balloon repels the like (same sign) charges in the wall and attracts the opposite charges. As a result, you have a net attraction between them even though the wall was originally neutral. The same happens in the glass on your TV screen. As it gets charged, it collects dust. Check it out.

FUN FACTS: SHOCKS

Static electricity can produce shocks when a charged body touches metal. The charge on your shoe soles induces static electrical charge on your body, and this charge, in turn, appears as a high voltage, which can be around 5,000 volts or even higher. (One volt is the energy required to transfer one unit of positive charge from its initial location to its final location.) As you touch a metal door knob or a filing cabinet, for example, there is a fast discharge, and you get a shock. A common complaint people have in the winter is that they shoot sparks when they touch objects. Their dry skin gathers a lot of positive electric charge as the skin gives up electrons to clothes made of polyester material, which attracts negative charge. To avoid such shocks, just wear all-cotton clothes, which are neutral. Also, moist skin drastically reduces the accumulation of charges. During dry summer days, moving cars can become electrically charged due to friction with the air. One way to discharge the static electricity is to use a metallic wire attached to the car's chassis, with one end touching the ground. Can you think of other ways to prevent or minimize electric shocks and nuisances arising from static electricity? In inkjet and laser printers, xerographic copiers, and electrostatic painting, just to mention a few applications, static electricity is vital. It allows a very precise and selective coating of paper surfaces with ink and toner. Also, car painting relies on electrostatic processes. Are you ready to find out new ways to turn static electricity into something useful?

2 | MAKING WATER DETOUR

See what happens when you bring an electrified balloon close to a stream of water.

SUPPLIES

- party balloon
- flexible plastic ruler, or piece of PVC pipe
- stream of water (from the tap in your kitchen sink)
- piece of flannel (clean and dry) or paper towel

SUPPLIES

- setup for Experiment 16, "Tubes of Light," under *Playing with Light: Optics*
- party balloon

STEP BY STEP

Open the tap so that a thin, steady stream of water flows out. Rub the balloon, the ruler, or the pipe against your hair, which should be clean and dry, or use the flannel (also clean and dry) or paper towel. Bring the balloon, ruler, or pipe close to the stream. What happens? Repeat the experiment using two balloons or two rulers instead of just one, with the stream in between.

A STEP FURTHER

How about guiding light?

FUN FACTS

The water molecules have a very special property. You can think of them as having two separated opposite charges. Molecules

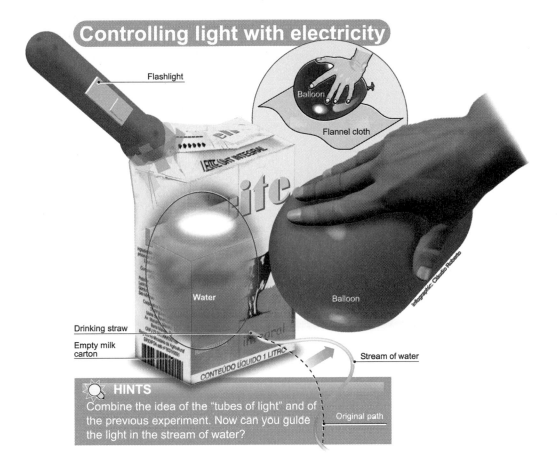

Controlling light with electricity

Flashlight

Balloon

Flannel cloth

Water

Balloon

Drinking straw

Empty milk carton

Stream of water

CONTEÚDO LÍQUIDO 1 LITRO

Infographic: Claudio Roberto

HINTS
Combine the idea of the "tubes of light" and of the previous experiment. Now can you guide the light in the stream of water?

Original path

with this property are called *polar*. When you bring a charged object close to the water, it attracts the opposite charge in the mobile water molecules more than it repels the charge with the same sign. The net result is attraction, as you can easily demonstrate by bending water toward you with the charged balloon. Once the stream of water is bent, light reflected in the inside walls of the stream will "go with the flow." You end up having electricity controlling the path of light.

Use a pipe to make a lamp light up.

SUPPLIES

- fluorescent lamp (even a damaged one will do, provided the glass is intact)
- piece of PVC pipe
- soft plastic (a plastic bag will do it)
- paper towel or piece of flannel

CAUTION

Be careful when handling fluorescent bulbs; the inner coating of the glass is toxic; adult supervision is recommended.

STEP BY STEP

Rub the PVC pipe with the paper towel, as indicated. In a dark room, bring the pipe close to the lamp and see what happens. You

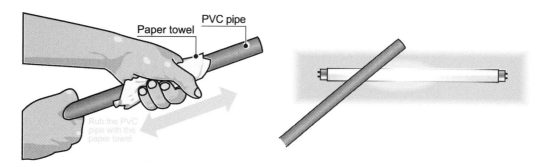

Paper towel — PVC pipe

Rub the PVC pipe with the paper towel

can also rub the fluorescent lamp directly using a soft plastic, moving the plastic back and forth along the lamp, faster and faster. (Hold one end of the lamp with one hand and rub it quickly with the other hand. Wear protective gloves in this case to avoid accidents.) Can you use this trick in the next blackout?

FUN FACTS

A fluorescent lamp consists of a sealed glass tube that contains a very small amount of mercury vapor and an inert gas, typically argon, kept under very low pressure. The tube is also coated with phosphor powder along the inside of the glass. When you rub the PVC pipe, it becomes negatively charged. If you rub the lamp directly with soft plastic, the lamp's glass tube becomes positively charged. In both cases, the electric charges accumulated in the glass wall produce an electric force strong enough to accelerate a few free electrons (with negative charge), which are always available inside the tube. These very few electrons can hit some mercury atoms and take additional electrons from them, triggering a cascade process that results in additional free electrons. These electrons, in their turn, hit neutral mercury atoms along their path and transfer energy to them. (The energy of the mercury atoms is increased, so they become "excited.") As the excited mercury atoms return to their relaxed state, they emit ultraviolet (UV) light, which is not visible. As the UV light hits the phosphor powder, the faint light you see is emitted either by approaching the charged pipe or rubbing the lamp with soft plastic. That's why such a soft glow is often called "phosphorescence."

4

SALT WATER TURNS INTO GAS: ELECTROLYSIS

You can transform water into oxygen and hydrogen using electricity.

SUPPLIES

- plastic bottle, 1 pint (500 ml)
- 2 pencils (you can cut a whole pencil in the middle to make two)
- superglue
- electric wiring
- 2 batteries (1.5 volts)
- party balloons
- salt water

STEP BY STEP

Carve the pencil points on both ends to expose the graphite inside, as shown in the figure. Make holes just the right size to fit the pencils in the side of the plastic bottle. Fit them in and use superglue to seal the holes. Fill the bottle with the salt water and fit the opening of the balloon over the top of the bottle, as shown. Twist the end of the wire around the graphite and attach the other end

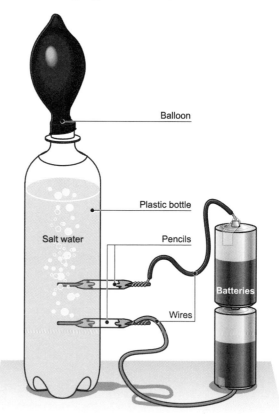

Balloon

Plastic bottle

Salt water

Pencils

Batteries

Wires

of the wires to the tops of the batteries (with tape), closing the circuit, as the picture shows. After 20 or 30 minutes, squeeze the balloon. What happens to the pencil points inside the water as time passes?

CAUTION ────────────────

The gas that fills the balloon is flammable. Be careful not to let the gas get near an open flame or other heat source.

FUN FACTS

Two hydrogen atoms (H) plus an oxygen atom (O) make up a water molecule (H_2O). Water molecules, which are neutral (have no net charge), can dissociate, producing a positive bit (H^+) and a negative bit (OH^-). (You may know that these bits are called *ions*.) Common salt (sodium chloride, or NaCl for short) is composed of an atom of sodium (Na) and one of chlorine (Cl). When dissolved in water, salt separates in a positive bit (Na^+) and a negative bit (Cl^-). The H^+ bits are attracted to the graphite connected to the negative end of the battery. There, pairs of bits become hydrogen gas as they receive two electrons from the battery, thus becoming neutral. As the H^+ disappears, the OH^- ions are left in the solution. The Cl^- bits are attracted to the graphite connected to the positive end of the battery. In this case, pairs of Cl^- bits deliver two electrons to the battery, forming chlorine gas, Cl_2. So, in the end, the batteries basically give and take equal amounts of electrons and you have Na^+ and OH^- ions as NaOH solution. This costs energy, of course.

There is something you can do to follow up on what is going on. Prepare a red cabbage juice. (Slice the red cabbage, put the slices in a blender with water, and blend them together. Filter the mixture with a sieve to get your juice.) The red cabbage juice is an "indicator," as chemists call it. If you pour some of the juice in vinegar, which is an acid, the vinegar becomes pink. If you pour some of the juice in water with soap, the soapy water becomes green. After the balloon is partially inflated, put some droplets of the juice in the bottle with salt water. What happens? Is NaOH solution acidlike or soaplike? You can also use your red cabbage juice to discover which liquids are acids and which ones are soaplike (or *bases*, as the chemists call them).

CAUTION ———————————————————

Even though only small quantities of hydrogen and chlorine gas are produced, avoid breathing the gases. Hydrogen is flammable, so do your experiment away from flames. The resulting solution of sodium hydroxide, or lye, is caustic, so avoid touching it. Stop your experiment as soon as you see the color changes in the red cabbage indicator.

———————————————————————

5

ELECTRIC GATES: THERMAL RELAYS

Interrupting electric currents with the heat of a flame.

SUPPLIES

- square piece of wood with two rectangular wood sticks fixed to it (see figure)
- battery (1.5 volts)
- small light bulb (for example, from a flashlight)
- connection wires
- aluminum foil
- piece of paper
- small flame (from a lighter or a match)
- masking tape
- glue

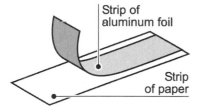

Strip of aluminum foil

Strip of paper

STEP BY STEP

Cut two equal rectangular pieces of paper and aluminum foil 1/2 x 1 3/4 in (1.5 x 4 cm). Glue the less reflective side of the aluminum foil to the piece of paper (your "relay"). Fix one end of two connecting wires to the wood sticks with masking tape, as indicated in the figure. Attach with masking tape one end of the relay (aluminum foil side down) to the connecting wire end fixed in one of the sticks. Let the other end of the relay just touch the other end of the connecting wire fixed on the other wood stick. Make sure that when the relay, lamp, and battery are connected, forming a closed circuit, the lamp is switched on. Bring a flame close to the free end of the relay from below, keeping it about 3/4 in (2 cm) from the relay. What happens?

FUN FACTS

The connection wires and the aluminum foil are very good conductors of electricity. (You can think of them as pipes with a large diameter that lets water flow much more easily than one with a smaller diameter, which you can associate with a very thin metallic wire.) As you close the electric circuit containing the lamp by connecting the batteries, an electric current consisting of billions of billions of "free" electrons circulates around the circuit. This multitude of electrons is driven by the

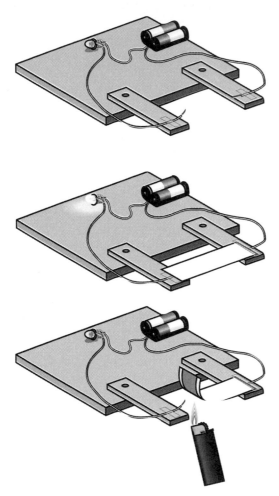

batteries. Without this driving force, the electrons would simply wander around randomly inside the piece of aluminum foil and in the wires and tungsten filament, since they have disordered kinetic energy associated with temperature (thermal energy). As the electrons cross the tungsten filament, they collide more often with the nuclei of the tungsten's atoms and make them vibrate harder, thus increasing the filament's temperature. As the filament is heated up, it emits light, just as black coal does in a bonfire. When the aluminum foil and paper glued together are heated up by the flame, they expand differently. (The aluminum foil, which is a metal, expands more than paper.) As a result, the relay curves up, interrupting the electric current crossing the lamp, so it stops emitting light. Thermal relays are used to limit electric currents in circuits. (If the current exceeds a certain value, the heat it generates in the relay makes it open, similarly to what happens in your thermal relay.) Now, can you figure out how to build other kinds of simple electric relays and relays to control other things, such as water and air flow?

6

ELECTRIC HOIST: ELECTROMAGNETS

Find out how magnetic hoists work.

SUPPLIES

- large screw or nail
- thin copper wiring, covered with plastic
- battery (1.5 volts)
- adhesive tape
- metal paper clips or tiny nails
- small pieces of paper or plastic (optional)

STEP BY STEP

Wrap the copper wiring in a spiral around the large nail or screw. Bring the nail or screw near to the pile of paperclips or tiny nails (try with other small objects). What happens? Strip the plastic insulation from the ends of the wires and attach them to the ends of the battery with adhesive tape. How do magnetic hoists work, then? Compare your hoist with a permanent magnet. What do the two have in common? Does your hoist attract paper or plastic? Or a magnet? What happens if you change the electrical connections of the hoist to the opposite ends of the battery?

CAUTION

The battery might become very warm because it is directly shorted. To avoid getting burned, hold the battery by a string attached or do the experiment for only a short time.

A STEP FURTHER

What do you think about replacing the hook in the "Hydraulic Robots" (Experiment 25 under *Fun with Mechanics*) with this electromagnetic hoist?

Electrifying Experiments: Electricity and Magnetism

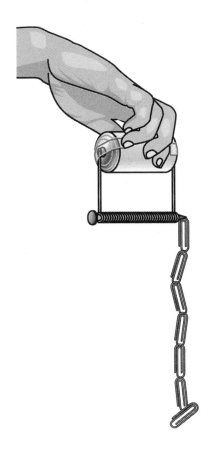

FUN FACTS

The electric hoist, as the proposed experiment allows you to demonstrate, acts as a magnet with a north pole (N) and a south pole (S). If you bring two magnets together with both their north poles facing each other, there is repulsion between them. The same happens if you bring their south poles together. If you bring the north pole of a magnet close to the south pole of another magnet, there is an attraction between them. (To demonstrate these effects, use small magnets to avoid injuring yourself. How do the attraction and repulsion forces vary with distance when the poles are closer?) Now, place your hoist close to a magnet. What happens if you invert the electric current passing through the hoist's wiring by changing the connections of the hoist with the battery ends (interchange + and –)? Does the direction of the current affect anything? You can also build another hoist and see how the two hoists react to each other.

7

CHAOTIC PENDULUM

Build a totally chaotic pendulum using magnets

SUPPLIES

- 1 donut-shaped magnet for the pendulum (for example, from an audio loudspeaker)
- 3 or more identical magnets or pieces of magnets
- piece of magnet-attracting metal

(the outside of the bottom of a large can, like those imported cookie tins, or its lid)
- 5 to 8 ft (1.5–2.5 m) of string
- 3 small nails
- sheet of paper or cloth (optional)

SPECIAL WARNINGS

While you are handling two magnets, avoid letting them get too close. They could suddenly attract each other and "bite" your hand. If two magnets stick together, you can separate them by sliding one up and the other down.

STEP BY STEP

Hang the pendulum magnet by three pieces of string, spaced equally apart on the magnet, and all three tied to a long piece of string for the handle (as shown). Tie the other end of the string to the ceiling or to a broom handle held up by two chairs, for example. Place on the floor, below the pendulum, the metal can lid with the other magnets or pieces evenly spaced on top of it. Adjust the length of the string as necessary to let the pendulum hang very close to the other magnets. Check if these magnets attract or repel the pendulum magnet and discover how to make the pendulum move chaotically. To give it the "magic touch," just cover the other magnets with a piece of paper or cloth. Let the pendulum swing normally and then bring the "chaos factor" onto the scene. Turn the magnets upside-down, transforming attraction into repulsion, and vice versa. Move and turn the lid the magnets are stuck to, to change their positions. Try to create total chaos with your new pendulum.

ALTERNATIVE

You could place the stuck magnets very close to each other so that there is a net repulsion or a net attraction between them and the pendulum magnet—this is analogous to the scattering of a neutron (pendulum magnet) by an atom (pendulum magnets). Neutrons are electrically neutral particles that interact with atomic magnets. Magnetism, at the atomic level, is associated with electrons. (An electric current in a loop is equivalent to a magnet, as the previous experiment demonstrates). Another possibility is to organize the magnets with their poles in alternation, like a chessboard, so that some produce attraction and others repulsion (called *antiferromagnet*). These alternatives can help you to visualize the interactions and arrangements that exist in the atomic world.

FUN FACTS

The pendulum without the magnets below it oscillates back and forth due to the gravitational attraction exerted by the Earth. (In null gravity, no oscillation would happen.) When you place the magnets just below the pendulum, additional magnetic forces act on the pendulum because the pendulum itself is a magnet and the magnets that are close to it can either attract or repel the pendulum. In this way, a regular motion can become chaotic.

8

PAINTING PICTURES WITH AN ELECTRIC HOIST

Turn an electric hoist into a cool paintbrush and have a ball.

SUPPLIES

- setup of Experiment 6, "Electric Hoist: Electromagnets"
- computer monitor screen (preferably white) or TV screen (in color)

STEP BY STEP

Turn on the computer monitor or TV. Hold the hoist (electromagnet) perpendicular to the screen, with its end near the screen. Turn the hoist on and off and see what happens. Then, hold the hoist touching the screen, parallel to it, as the illustration shows, and activate it again. Try using two batteries (in series) instead of just one in the hoist and try holding it at various angles in relation to the screen. What happens? How did you manage to turn the hoist into a paintbrush?

CAUTION ——————————————

The battery might become very warm because it is directly shorted. To avoid getting burned, do the experiment for only a short time.

A STEP FURTHER

See if this trick works with a liquid crystal screen (such as many laptop computers and digital watches have).

FUN FACTS

The image formed on the screen of most modern computer monitors and televisions is produced by beams of electrons that sweep across the screen all the time, from left to right and top to bottom. Each electron that hits the phosphorus-coated screen creates a spark of light, and we perceive all the sparks together as a complete image. The electron in movement, in the same way as an electric current, is affected by the magnetic field of an electromagnet (see Experiment 9, "Electric Motor," next). Different orientations of the hoist in relation to the screen correspond to different orientations of the magnetic field it generates. Which orientation most affects the beam of electrons that reach the screen?

The experiment proposed could, in principle, be performed with a permanent magnet, like the kind found in audio loudspeakers. This, however, would run the risk of leaving marks or spots on the screen, which could remain even after the appliance is turned off. The use of the electric hoist eliminates this risk.

9

ELECTRIC MOTOR

Build the world's simplest electric motors.

Electrifying Experiments: Electricity and Magnetism

SUPPLIES

- 1 yd (1 m) copper wiring with insulation, about 1/5 in (0.51 mm) in diameter
- 2 pieces of thick wire, each 8 in (20 cm) long
- battery (1.5 volts)
- magnet, such as from a speaker (see illustration)
- masking tape
- metal file or steel wool

STEP BY STEP

To make the motor's coil, wrap the copper wire around the battery. Your 1 yd (1 m) of wire should make 22 to 25 turns around the battery with some left over. Keep 3/4 in (2 cm) from each end of the wire for the motor's shaft or axis (see the illustration). Make two holders for the coil with the thick wire, bending it as shown in the design, removing any rust that may be present with the metal file or steel wool. Attach the wire holders for the coil to the two ends of the battery, as the picture shows, using masking tape to hold them firmly in place. Strip the entire

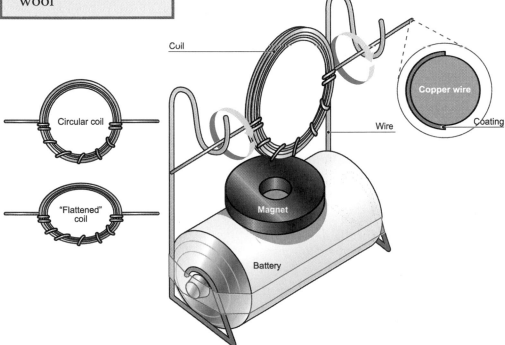

Circular coil

"Flattened" coil

Coil

Copper wire

Wire

Coating

Magnet

Battery

insulating coating from the copper wiring of one end of the motor's shaft and only half the insulation from one half—along one side—of the other (see the inset illustration). Place the coil on its holders and the magnet on top of the battery (see the picture). If necessary, gently spin the coil to jump-start the motor.

A STEP FURTHER

Flatten the coil as shown and see how its new geometry affects the motor's efficiency. What relationship can you see between the two coils and the "Ballerina's Trick" (Experiment 35 under *Fun with Mechanics*)? You can use a TV screen or computer monitor to follow the coil's movements (see Experiment 17, "Slow-Motion Camera," under *Playing with Light: Optics*).

A STEP EVEN FURTHER: LET'S SPARE A MAGNET! ★★

What if you replace the magnet of the previous model with an electric hoist?

SUPPLIES

- copper wiring with insulation about 7.7 yd (7 m) long and 1.2 mm diameter
- 4 in (10 cm) piece of the same copper wiring
- copper wiring with insulation, 2.2 yd (2 m) long and 0.51 mm in diameter
- 2 or 3 batteries, 1.5 volt
- small (AA) battery or thick pencil
- piece of PVC pipe around 1 in (2.5 cm) diameter with a small cut along its axis to hold one end of the copper wiring
- masking tape
- 2 small pieces of rigid plastic

To make the motor's coil, wrap the 2.2 yd (2 m) of copper wiring around a small battery (AA) or a thick pencil. Then follow the same steps used in the previous model, keeping 3/4 in (2 cm) from each end of the wiring for the motor's shaft or axis. Make sure that the entire insulating coating is removed from the copper wiring of one end of the motor's shaft

and only half the insulation from one half—along one side—of the other. Strip the entire insulating coating from both ends of the 7.7 yd (7 m) copper wiring and do the same with the 4 in (10 cm) piece of the same wiring. Then wrap the 7.7 yd (7 m) wiring around the piece of PVC pipe, sparing 4 in (10 cm) on both ends. The coil thus obtained corresponds to the electric hoist, which will replace the magnet of the previous model. Use one of the coil's ends to fix it and to produce a needle-like holder for one end of the motor's shaft. The other end of the coil should be attached to one pole of the batteries. Make a similar holder for the other motor's shaft with one end of the 4 in (10 cm) piece of wiring. Fix the holder to the coil with a tie, as indicated in the figure. The other end of the holder's wiring should be attached to the other pole of the batteries. Get one end of the motor's shaft stuck into each piece of rigid plastic with the needle hole in between to prevent the shaft from moving sideways.

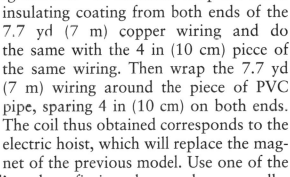

Electrifying Experiments: Electricity and Magnetism

FUN FACTS

The coil (rotor) is just a variant of the electric hoist, so when an electric current passes through its wiring, it acts like a magnet. Two magnets put close to each other get stuck together. That's inevitable. It is just like objects that tend to stay as close as possible to the Earth's surface due to gravity. (By the way, the Earth itself turns out to be a big magnet, as you can demonstrate with a magnetic needle.) The motor stops when either the south pole of the motor's moving coil and the north pole of the magnet (first model) or of the "hoist" (second model), or vice versa, face each other. That's why you have to strip only half the insulation from one shaft end of the motor. The moving coil is no longer a magnet when no current passes through its wiring. The coil continues spinning due to its inertia (conservation of angular momentum), because nothing prevents it from doing so (see Experiment 36, "The Bicycle's Trick," under *Fun with Mechanics*). Now, can you think of a way to keep the coil rotating without having to interrupt the current flow in the coil? Will this enable the rotor of your new model to spin faster? Can you also make your electric motor work in reverse, as a generator, so that it produces electricity instead of consuming it?

HINT

Attach a home-made turbine to the rotor to keep it turning as a water stream or a vapor jet hits the turbine's blades; see Experiment 4, "Steam Machine," under *The World of Atoms and Our World: Cold, Heat, and Giant Bubbles.* You can replace the battery with an LED or a flashlight lamp.

10

CRAZY TOBOGGAN:
ELECTROMAGNETIC BRAKING

What about making a toboggan unique?

STEP BY STEP

Bend the piece of cardboard into a track with a surface 1 3/4 in (4.5 cm) wide and side walls 3/4 in (2 cm) high (see illustration). When bent, the cardboard becomes more resistant (see Experiment 2, "How the Weak Become Strong," under *Fun with Mechanics*). Stick the track on the bottom of the pot or cookie sheet with the tape, as the picture shows. To make a descending ramp for the toboggan, just tilt the pot. Place the magnet on the top of the toboggan track and let it slide down. What happens when the magnet reaches the part of the track in contact with the aluminum pot? To make this experiment even more fun, you could glue the magnet to a little plastic toy car. Check out what it can do as it races down the track.

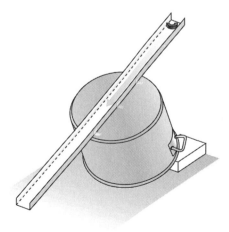

A STEP FURTHER

Take a piece of aluminum foil, about 4 x 12 in (10 cm x 30 cm) and make a ramp with it. What happens when the magnet slides down this ramp? What is different in this case?

FUN FACTS

A magnet attracts metal (steel) paper clips, but not aluminum. However, as a metal, aluminum can conduct electricity. This means that it has lots of free electrons, which are very mobile. The magnet is not static. It moves down the track due to gravity. The metal bottom perceives a magnetic field varying in time as the magnet passes by. It is not happy about that. It prefers to have no change at all in the magnetic field, so the free electrons in the metal bottom circulate, producing "eddy currents" (also called *Foucault currents*). These currents, in turn, create a magnetic field to balance the varying magnetic field as the magnet passes by. The opposing magnetic field thus produced is equivalent to a regular magnet. In fact, the hoist (see Experiment 6, "Electric Hoist: Electromagnets," in this part of the book) shows that an electric current can have the same effect as a magnet. In the case of a strip of aluminum foil, the Foucault currents are very weak, since the area of the foil's cross section (its thickness times the width of the strip) is very small. Consider the following analogy. A pipe with a large diameter lets water flow much more easily than one with a small diameter. In the same way, a thick electric wire offers less resistance to the passage of electricity than a very thin wire. Now compare a thick piece of aluminum with a very thin one.

What can you conclude? The Foucault currents are also present in the electric transformers on electric poles. Transformers are made of thin plates of metal, coated with insulating enamel and then placed together. If a single block of material was used, the Foucault currents would come into play strongly, undesirably consuming electrical energy. These currents are in fact used to measure how much electricity is consumed in your home. Some meters of electric energy use an aluminum wheel, which spins as electricity is consumed. It has small electromagnets that generate Foucault currents and make it spin, which in turn causes the meter to function, measuring how much electricity has been used.

11

MAGNETIC LEVITATION

Discover how to make objects levitate, using magnets.

SUPPLIES

- 2 donut-shaped magnets (for example, from a speaker no one uses; see illustration)
- 2 small ring-shaped magnets (for example, from a pair of headphones no one uses)
- double-pointed pencil
- 4 nails
- barrier (for example, a drinking glass)

SPECIAL WARNINGS

When handling the two larger magnets, be careful not to let them get too close to each other. They could suddenly attract each other and "bite" your hand. If they get stuck together, gradually slide one up and the other down, until they separate.

STEP BY STEP

Place the small ring magnets on both ends of the double-ended pencil, as the picture shows, pressing them with a little force so they stay on the pencil ends. Place the large donut-shaped magnets on their edges on the table. To keep them from rolling away, you can place two nails under each one, as shown. Place the pencil point with its ring magnet inside the large magnet's hole. If it doesn't levitate, turn the large magnet around and try again. To stay suspended, the ring magnet should be slightly off center in relation to the large magnet with the barrier holding the pencil in place, as the detail on the picture shows. (At this stage, you will be holding one end of the pencil with your hand.) Bring the barrier close to prevent the pencil from escaping. Now, place the second large magnet as shown. The ring magnet stuck on the free end of the pencil should again be slightly off center (see the picture). If this magnet doesn't levitate, turn the large magnet around and try again. It might be necessary to make small adjustments in the positions of the magnets and the barrier. Try spinning the pencil while it is suspended. What if you took away the barrier? Would the pencil still levitate?

FUN FACTS

Trains that levitate are based on the same principle of magnetic repulsion explored in this experiment. However, magnetic levitation involving permanent magnets, such as those used in loudspeakers, is inherently and unavoidably unstable. Although two magnets can repel each other, a permanent magnet suspended above another permanent magnet will *always* crash (you can check that easily with small magnets). The equilibrium in this case is unstable, like a lightweight ball on top of a hill. A slight lateral perturbation (like a mild breeze) can make the ball roll down the hill and never return. In the case of levitating trains, electromagnets are used (see Experiment 6, "Electric Hoist: Electromagnet") instead of permanent magnets. In some models of levitating trains, the electric currents that circulate in electromagnets are controlled by sensors installed in the train cars. The currents are continually adjusted so that the train can move without "leaving the rails." Thanks to this trick, trains don't need a barrier to keep them suspended. Another possibility is magnetoelectrodynamic levitation (induced magnetism in a conducting track, created by a very fast moving (over 100 mph) magnetized train). In this case, the electric current circulating in the "electric hoists" carried inside the train is alternating current (AC). This means that the current keeps changing its direction. (A 60 Hz alternating current, for example, changes its direction 60 times per second.) The changing direction of the current creates an alternating magnetic field that half of the time points in one direction and half of the time points in the opposite direction. (You can do that with your hoist by switching the connecting wires back and forth between the + and − terminals of the battery.) This changing magnetic field generates alternating currents in the metallic track underneath the train (see previous experiment.) The track thus acts as a dynamic magnet that counteracts the alternating magnetic field of the train. Since the magnetic field produced by the track tends to cancel the train's magnetic field, there is a net repulsion and the train levitates nicely.

It's really easy to mute a radio playing at maximum volume.

SUPPLIES

- battery-operated radio
- cardboard box with a lid (like a shoebox)
- empty juice carton lined with aluminum foil
- aluminum foil
- microwave oven

STEP BY STEP

Tune the radio to a station with a strong signal at maximum volume (with adult supervision). Now, place it inside the shoebox. With the lid on, can you still hear the radio? Next, open the juice carton enough to fit the radio inside and close it up again, as shown (with adhesive tape, if necessary). What happens to the radio?

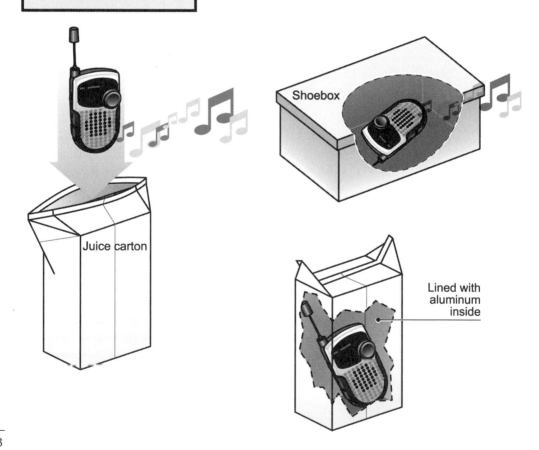

Juice carton

Shoebox

Lined with aluminum inside

ALTERNATIVE

Place the radio inside a metal cooking pot (with a metal lid), or wrap it (including the antenna) in aluminum foil and see what happens. Repeat the experiment with a cellular phone inside a metal pot (lid on) or wrapped in foil. Does it ring when you call its number? You can also wrap aluminum foil around a remote control (to TV, VCR, garage door, or car alarm) and see if these devices still work. What do these remote controls, the radio, and the cell phone all have in common?

A STEP FURTHER

Fill a cup with water and pour an equal amount into the empty juice carton (lined with aluminum). Place them both in a microwave and turn it on for about 30 seconds. Compare the water in the cup with that inside the carton. What is the connection between this experiment and the silent radio experiment?

FUN FACTS

A metal pot or aluminum foil has lots of "free electrons," which can move around inside the metal walls. These electrons act like tiny antennas as an electromagnetic wave hits the metal's surface. The wave's electric field pushes free electrons and causes them to accelerate. These accelerating electrons redirect (absorb and reemit) the wave in a new direction, producing a mirror reflection. Just as a box made of metal mirrors keeps visible light out, a box made with metal walls keeps radio waves and the microwaves of a cell phone out. The same holds when the electromagnetic wave is generated inside a metal box. The wave becomes trapped inside it, as it is reflected by the metal walls. That's exactly what happens in a microwave oven. (The holes in the metal grid on the oven's glass window are smaller than the microwave wavelength, so the grid also reflects the waves back.) When you turn on a radio or receive a call on a cell phone, external electromagnetic signals activate these devices. When they are inside a metal pot (lid on) or wrapped in aluminum foil, the free electrons in the pot or in the foil respond quickly, reflecting back the incoming signals. This does not happen with the shoebox, since cardboard is an insulator and does not have free electrons. (Try using strips of paper or cardboard to connect a small lamp from a flashlight, for example, to a battery and see if the lamp lights up. You can also try wetting the paper or cardboard. Try wetting it with salt water. What happens then?)

13 CAR CONTROL VERSUS TV CONTROL

Have you ever compared the several remote controls you have at hand?

SUPPLIES

- car remote control and TV remote control
- metal pot (big enough for both controls)

STEP BY STEP

Place the TV remote control on the bottom of the pot and try to switch your TV on as usual. Does the control need repair? Now, do the same with the car remote control and see if the car responds to your command.

A STEP FURTHER

Direct your controls at surfaces so that by reflection the TV or car should respond. Repeat the experiment with the controls inside the pot.

FUN FACTS

Car remote controls use radio waves whose wavelength is comparable to the dimensions of your pot. The signal goes around the corner (the pot's edge) and arrives at the car's antenna. Obstacles that are small compared to the wavelength of light allow the wave to continue traveling (see Experiment 13, "New Discoveries with Polaroids," under *Playing with Light: Optics*). That's why radio waves are so effective for transmitting radio signals. TV remote controls, on the other hand, use infrared light, whose wavelength is much shorter compared to the dimensions of ordinary objects. In this case, the infrared signal is confined to the pot. You can still get the signal reflected back up by the pot's bottom as if it were a flat mirror. (Try it with the pot opening placed in front of your TV and then move the pot so that the signal can hit the TV infrared receiver.)

Patterns for Fun with Mechanics, Experiment 13:
The Square Wheel and Others

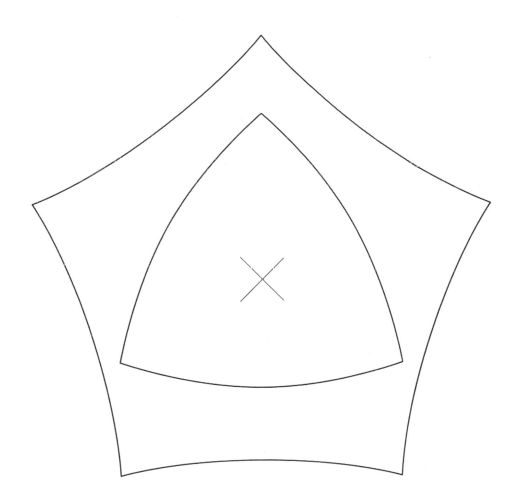

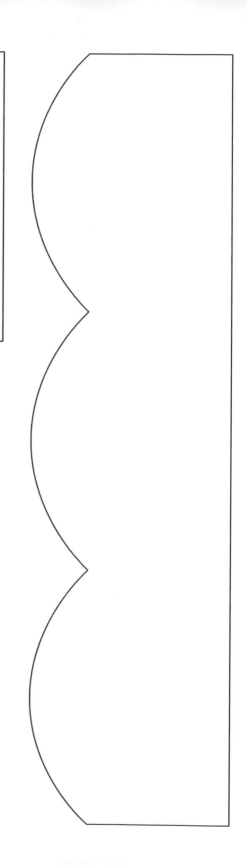

Patterns for Playing with Light: Optics, Experiment 3:
Challenge Your Perception

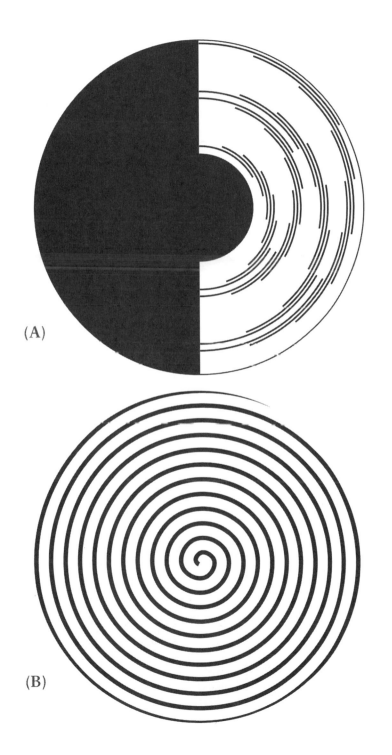

(A)

(B)

Paint the sectors of Newton's disk (pattern C) according to the instructions in the text.

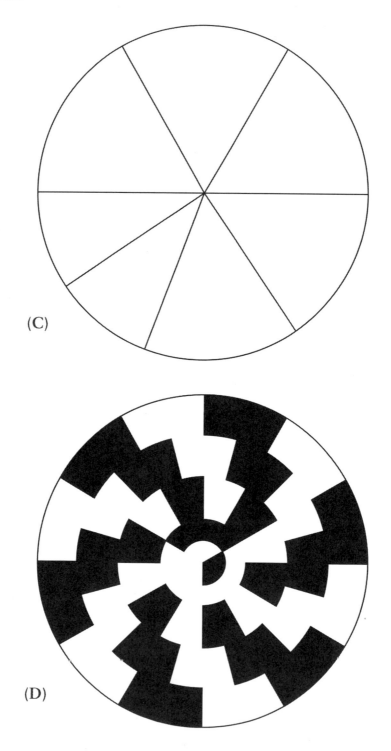

(C)

(D)

Index

http://www.phptr.com/

Prentice Hall PTR   InformIT   InformIT Online Books   Financial Times Prentice Hall   ft.com   PTG Interactive   Reuters

TOMORROW'S SOLUTIONS FOR TODAY'S PROFESSIONALS

Prentice Hall **Professional Technical Reference**

Browse | Book Series | What's New | User Groups | Alliances | Special Sales | Contact Us

Search | Help | Home

Quick Search

PTR Favorites

Find a Bookstore

Book Series

Special Interests

Newsletters

Press Room

International

Best Sellers

Solutions Beyond the Book

Shopping Bag

Keep Up to Date with

PH PTR Online

We strive to stay on the cutting edge of what's happening in professional computer science and engineering. Here's a bit of what you'll find when you stop by **www.phptr.com**:

What's new at PHPTR? We don't just publish books for the professional community, we're a part of it. Check out our convention schedule, keep up with your favorite authors, and get the latest reviews and press releases on topics of interest to you.

Special interest areas offering our latest books, book series, features of the month, related links, and other useful information to help you get the job done.

User Groups Prentice Hall Professional Technical Reference's User Group Program helps volunteer, not-for-profit user groups provide their members with training and information about cutting-edge technology.

Companion Websites Our Companion Websites provide valuable solutions beyond the book. Here you can download the source code, get updates and corrections, chat with other users and the author about the book, or discover links to other websites on this topic.

Need to find a bookstore? Chances are, there's a bookseller near you that carries a broad selection of PTR titles. Locate a Magnet bookstore near you at www.phptr.com.

Subscribe today! Join PHPTR's monthly email newsletter! Want to be kept up-to-date on your area of interest? Choose a targeted category on our website, and we'll keep you informed of the latest PHPTR products, author events, reviews and conferences in your interest area.

Visit our mailroom to subscribe today! **http://www.phptr.com/mail_lists**